1961.1962.1963.1964.1965.1966.1967.1968
1974.1975.1976.1977.1978.1979.1980.1981
1987.1988.1989.1990.1991.1992.1993.1994
2000.2001.2002.2003.2004.2005.1931.1932
1938.1939.1940.1941.1942.1943.1944.1945
1951.1952.1953.1954.1955.1956.1957.1958
1964.1965.1966.1967.1968.1969.1970.1971
1977.1978.1979.1980.1981.1982.1983.1984
1990.1991.1992.1993.1994.1995.1996.1997
2003.2004.2005.1931.1932.1933.1934.1935
1941.1942.1943.1944.1945.1946.1947.1948
1954.1955.1956.1957.1958.1959.1960.1961
1967.1968.1969.1970.1971.1972.1973.1974

Reflections
from the Frontiers

EXPLORATIONS FOR THE FUTURE   Gordon Research Conferences, 1931–2006

Editors

Arthur A. Daemmrich

Nancy Ryan Gray

Leah Shaper

Gordon
Research
Conferences

CHEMICAL
HERITAGE
FOUNDATION

Philadelphia, Pennsylvania

Contents

REFLECTIONS ON MATTER

Everyone Has a GRC *Story*

I believe everyone who has attended a Gordon Research Conference (GRC) has a "GRC story." This commemorative publication, celebrating the seventy-fifth anniversary of the Gordon Research Conferences, is a testament to that belief.

Nancy Ryan Gray
Director
Gordon Research Conferences

There is the GRC story itself, the history of a unique organization that grew from humble beginnings and ultimately set the standard of excellence against which other conferences are compared. There is the story of GRC's founder, Neil Elbridge Gordon, and his creation of a simple venue, free from distraction, in which he and his colleagues could discuss and debate the frontiers of scientific research. There are the stories of formalizing a framework for supporting the conferences; the visionary leadership provided by GRC's governance and Neil Gordon's successors (my predecessors); and the remarkable growth of the conferences in scientific breadth, number, and geographic distribution. But most significant are the stories of those who have made GRC the enduring organization it has become. These are the stories of the conferees themselves: the chairs, vice chairs, discussion leaders, and speakers; the first-time attendees and longtime loyalists; the scientifically famous; and those unforgettable conferees (and you know who you are) who have become infamous.

I have always been a sucker for a good story. As a result this commemorative publication was conceived as a means to celebrate the seventy-fifth anniversary of GRC by capturing a small, but significant, subset of GRC stories. The contributors were asked to reflect on how their GRC experiences affected both their careers and their specific scientific disciplines. What could they tell us about how GRC pioneered new areas of research, started important collaborations, launched journals and professional societies, or contributed to breakthrough technology and innovations? Why did they keep coming back? What is it about the GRC experience that makes it unique?

My GRC story begins in the summer of 1983 when I was a graduate student at Penn State. My adviser, noted organic geochemist Peter H. Given, informed me that he was sending me to New Hampshire for a week to attend the Analytical Pyrolysis Gordon Conference. The conference was chaired by Fred Shafizadeh of the University of Montana, who was considered the world's leading expert on the pyrolysis of wood. In preparation for the conference I had assembled a large stack of publications related to the field. These articles chronicled the debates over pyrolysis mechanisms and theory, as well as the utility of pyrolysis for characterizing synthetic polymers and large biopolymers. Although I expected the conference program to address these subjects, I was speechless to find myself sharing dinner with the primary authors on the mastheads. These were the gods of the discipline (Henk Meuzelaar, Hans-Rolf Schulten, Piet Kistemaker, Stan Israel, Bob Lattimer, and Kent Voorhees), and they were asking me about my thesis research and to please pass the salt.

During the poster session one evening I met two Dutch scientists (Jaap Boon, of the FOM Institute for Atomic and Molecular Physics, and Jan de Leeuw, of Delft University of Technology) presenting results from their analysis of peat samples from the Netherlands. Since I was studying peat samples from the United States, I spent most of the evening asking questions about using pyrolysis to study the recalcitrant portion of sedimentary organic matter. Two weeks later, literally hip deep in the swamps of the Florida Everglades gathering additional peat cores, I received a phone call from my thesis adviser. Dr. Given had been visited by the two Dutch scientists prior to their return to the Netherlands. The three of them collectively agreed to send me to Amsterdam where I would complete the analytical portion of my thesis research using state-of-the-art equipment at the FOM Institute. Going to New Hampshire for a week landed me in Europe for a year.

I continued to use analytical pyrolysis in my research after completing my Ph.D. and accepting a position at Exxon Production Research Company in Houston. I became a "regular" at the Analytical Pyrolysis GRC and was honored to be invited as a speaker for the 1989 and 1991 meetings. My career path took an unexpected turn in March 1989, however, when a big ship of Exxon's hit a big rock in Alaska (otherwise known as the *Exxon-Valdez* disaster). I transferred from fundamental research in oil exploration to a new environmental research group working on oil production. By 1993, owing to declining attendance, the Analytical Pyrolysis Gordon Conference was canceled.

In 1994 I left Exxon and my research career to accept a position with the American Chemical Society (ACS). Although I missed research, I enjoyed working for an organization dedicated to serving the chemical community, and I continued to see many of my GRC buddies at ACS meetings. In 2003, while writing a business plan for a new series of ACS conferences, I stumbled across the announcement that a new director was being sought to head GRC. It was an opportunity to write an amazing ending to my GRC story—and I'm a sucker for a good story.

I am extremely honored to serve as only the fifth GRC director in seventy-five years and to continue the tradition of excellence that makes the Gordon Research Conferences so successful. When asked what I admire so much about GRC, I reply, "It's the story." It's the story of an amazing organization conceived, built, and maintained by scientists, and the story of an organization that has done the same thing for three-quarters of a century and has done it exceedingly well. It is a fascinating story. It is a story that, as a scientist, makes me proud. And it is a story that continues to unfold.

seventy-five years of ACHIEVEMENT

1931

■ Summer research conferences on chemical physics are held at Johns Hopkins University.

■ Ernst Ruska and Max Knoll develop a prototype of the electron microscope.

■ Harold Urey proves the existence of deuterium.

1932

■ Five summer research conferences meet at Johns Hopkins.

1934

■ Conferences move to Gibson Island, Maryland.

■ Dorothy Crowfoot Hodgkin and John D. Bernal record the X-ray diffraction pattern of a protein.

1935

■ Wendell Meredith Stanley crystallizes tobacco mosaic virus, laying precedent for the molecular treatment of viruses.

1936

■ Alan Turing introduces the concept of the Turing machine to provide a mathematically precise definition of *algorithm*.

1937

■ Conferences are held at Virginia Beach.

■ The National Cancer Institute is established.

1938

■ Neil Gordon ensures continuity for the conferences by persuading the American Association for the Advancement of Science (AAAS) to manage the conferences; Gordon serves as conference secretary for the 1938 AAAS Special Research Conferences on Chemistry.

■ DuPont announces that it will start producing nylon, the world's first synthetic fiber, which was invented by Wallace Carothers in 1935.

1939

■ Gordon serves as secretary for the conferences and is named director in the fall.

1938, Newspapers report in 1938 that DuPont will begin production of nylon

1931, Harold Urey

30s

1946, ENIAC

1940

- Edwin McMillan and Phillip Abelson synthesize neptunium, the first transuranic element.

1941

- In order to expand meeting and lodging facilities, Gordon purchases the Symington House and Annex with $8,000 secured from AAAS and $1,000 from each of thirty-three companies with significant R&D operations. For many years afterward industrial sponsors are guaranteed conference registration slots in return for their generosity. The Symington House is renamed the Neil E. Gordon House by the AAAS executive committee.

1942

- Conferences are renamed the AAAS–Gibson Island Research Conferences.

- Enrico Fermi and his colleagues achieve a controlled, self-sustaining nuclear fission reaction.

- George Nicholas Papanicolaou introduces Papanicolaou's stain for smear preparations of bodily secretions, now commonly used for cancer screenings.

1943

- Los Alamos National Laboratory is founded, and Oak Ridge National Laboratory is built.

1944

- Streptomycin, the first antibiotic remedy for tuberculosis, is developed.

1945

- Gordon resigns; Sumner B. Twiss manages the conferences.

1946

- Twiss resigns; George Calingaert is appointed chair of the management committee.

- Conferences move to Colby Junior College in New London, New Hampshire.

- Argonne National Laboratory is founded.

- The world's first fully electronic, programmable, digital computer, ENIAC (Electronic Numerical Integrator and Computer), is developed at the Moore School of Electrical Engineering at the University of Pennsylvania under contract with the Army Ordinance Department.

1947

- W. George Parks is appointed director of the conferences, and headquarters move to Rhode Island State College (changed to the University of Rhode Island in 1951). Alexander M. Cruickshank is hired as assistant to the director, and Irene Cruickshank is hired as secretary and treasurer.

- The conferences are renamed the Chemical Research Conferences.

- John Bardeen and Walter Brattain invent the point contact transistor under the leadership of William Shockley at Bell Telephone Laboratories.

- Brookhaven National Laboratory is founded.

1948

- The Chemical Research Conferences are formally renamed the Gordon Research Conferences.

1949

- Percy Julian synthesizes Reichstein's Substance S, which is present in the adrenal cortex and differs from cortisone by only one specifically positioned oxygen atom. Hydrocortisone is produced from this substance today.

1949, Percy Julian

40s

1953, DNA double helix

1960

■ Theodore Maiman of Hughes Research Laboratories invents the laser (light amplification by stimulated emission radiation) using a ruby cylinder.

1950

■ The New Hampton School in New Hampton, New Hampshire, is added as a second conference site.

■ The National Science Foundation is established.

1961

■ The Tilton School in Tilton, New Hampshire, is added as the fourth conference site.

■ Marshall Nirenberg and Heinrich J. Matthaei decipher the mechanism of RNA transmission of DNA "messages" used to direct amino acids in protein production.

1951

■ Carl Djerassi's group at Syntex achieves the total synthesis of cortisone, as well as the first synthesis of norethindrone, the basis for the first oral contraceptive.

1953

■ James Watson and Francis Crick elucidate DNA's double helical structure.

1956

■ GRC incorporates as a nonprofit organization, establishing a board of trustees, and celebrates its twenty-fifth anniversary at the Hotel Commodore in New York City.

■ Dorothy Crowfoot Hodgkin determines the structure of vitamin B_{12}.

1954

■ The Kimball Union Academy in Meriden, New Hampshire, is added as a third conference site.

■ The world's first nuclear power plant in Obninsk, Kaluga Oblast, USSR, begins producing electricity for commercial use.

1952

■ Jonas Salk of the University of Pittsburgh develops a successful polio vaccine.

1957

■ The launch of *Sputnik 1* starts the space race between the United States and the USSR.

■ Polycarbonate plastics are developed for use in headlights, microwave cookware, and other high-impact applications.

1962

■ Lycra synthetic elastomeric fiber is first commercialized by DuPont. Lycra is a segmented polyurethane incorporated with other fibers to produce fabrics with elasticity.

1958

■ The GRC Selection and Scheduling Committee is formed.

SALK'S VACCINE WORKS!

POLIO VACCINE IS 'SAFE EFFECTIVE AND POTENT'

Way Out, Salk Report Shows: Enabled to Treat 50% More

Courtesy March of Dimes

1966, Stephanie Kwolek and Kevlar products

50s

60s

1970

- The liquid crystal display (LCD) is patented.

- Congress adopts the Clean Air Act, and President Richard M. Nixon establishes the Environmental Protection Agency.

1963

- The first winter conferences are held in Santa Barbara, California.

- Murray Gell-Mann proposes that protons, neutrons, and other subatomic particles are made up of quarks.

1966

- Conferences are held in Washington State.

- Kevlar is patented by Stephanie Kwolek of DuPont. Kevlar is an extremely strong, flexible, and flame-resistant synthetic organic fiber that is lighter than steel; it is used to make bulletproof vests, underwater cables, brake linings, and numerous other applications.

1971

- Conferences are held in Wisconsin.

- Intel develops the first general-purpose, programmable micro-processor, the 4004.

1977

- Raymond Damadian introduces magnetic resonance imaging (MRI) for medical diagnoses.

- Genentech produces somatostatin, the first human protein to be manu-factured in bacteria.

- Alan Heeger, Alan MacDiarmid, and Hideki Shirakama discover that doping organic polymers with atoms and small molecules facilitates their conductivity. A huge array of appli-cations of the resulting conducting polymers, such as rechargeable bat-teries, corrosion inhibition, antistatic dispersion, and flexible transistors, are still being explored today.

- Apple Computer releases the Apple II personal computer.

1964

- Proctor Academy in Andover becomes the fifth New Hampshire site.

1973

- Herbert Boyer and Stanley Cohen introduce recombinant DNA tech-nology.

1965

- Gordon Moore states that integrated circuits will double in complexity every 18 months. This becomes known as Moore's law, and with revisions, it becomes a truism for the next four decades.

1967

- Arthur Kornberg, Mehran Goulian, and Robert Sinsheimer synthesize a biologically active viral DNA molecule.

1975

- The first catalytic converters are installed in automobiles in response to motor-vehicle emissions regulations issued by Congress.

1978

- Genentech and the City of Hope National Medical Center announce the successful laboratory production of human insulin using recombinant-DNA technology.

1968

- Cruickshank is appointed director after Parks resigns.

1969

- The *Apollo 11* spacecraft lands on the Moon.

1977, Apple II

70s

1980

■ In *Diamond v. Chakrabarty*, the U.S. Supreme Court rules that genetically engineered bacteria are patentable.

1983

■ Kary Mullis and others at Cetus Corporation invent the polymerase chain reaction (PCR), which makes possible the replication of DNA sequences without the use of microorganisms in amounts large enough for analysis.

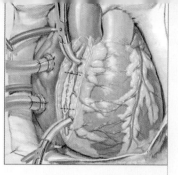

1988, Angina pectoris

1990

■ The first conferences are held outside the United States in Volterra, Italy.

■ The Human Genome Project is launched and later completed in 2003.

1985

■ The Antarctic ozone hole is confirmed (first identified by British Antarctic Survey in the 1970s).

■ Buckminsterfullerene (carbon-60), the only pure form of carbon aside from diamond and graphite, is discovered by Robert F. Curl, Jr., Harry W. Kroto, and Richard E. Smalley and named after the American architect Buckminster Fuller. Buckyballs are hollow round clusters of sixty carbon atoms.

1987

■ Salve Regina University in Newport, Rhode Island, is established as a summer site.

1988

■ Sir James Black receives the Nobel Prize in physiology or medicine for the development of propranolol, a beta-blocker that revolutionized the treatment of angina pectoris.

1991

■ Conferences are held in Germany.

1992

■ Conferences are held in Hawaii.

■ The FDA approves the use of Taxol (paclitaxel), a low-toxicity chemotherapy drug identical to a compound found in the bark of the Pacific yew tree, for the treatment of ovarian cancer.

1981

■ GRC celebrates its fiftieth anniversary with the production of *Gordon Research Conferences, Frontiers of Science; 50th Anniversary*. A special council meeting is held on 31 March at the Atlanta Hilton Hotel.

■ The scanning tunneling microscope is invented by Gerd Binnig and Heinrich Rohrer.

1982

■ Compact disc players are commercially available for the first time.

1993

■ Carlyle B. Storm is appointed director after Cruickshank retires. The Alexander M. Cruickshank Lectures are established to showcase leadership in biology, chemistry, and physics.

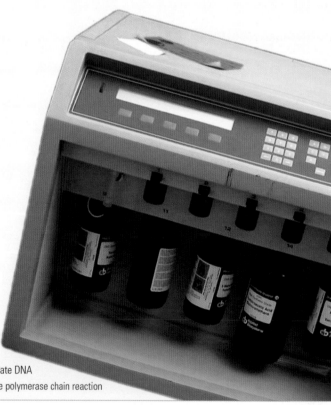

1983, Applied Biosystems Model 391 PCR-Mate DNA Synthesizer used in the first step of the polymerase chain reaction

80s

90s

1994

- Public access to the World Wide Web through dial-up services becomes available.

1995

- Conferences are held in Switzerland.

- Paul J. Crutzen, Mario J. Molina, and F. Sherwood Rowland are awarded the Nobel Prize in chemistry for their work in understanding the formation and decomposition of ozone.

1996

- Conferences are held in England, the Czech Republic, and Japan.

1997

- GRC celebrates its 1947 move to New England with the publication of *Gordon Research Conferences: 50 Years in New Hampshire*. A special council meeting is held on 8 August at the Ramada Logan Hotel in Boston.

- Conferences are held in France.

- Half the Nobel Prize in chemistry is awarded to Paul D. Boyer and John E. Walker for understanding the mechanism and structure of adenosine triphosphate (ATP), the principal carrier of chemical energy in the body. The other half is awarded to Jens C. Skou for the discovery of the enzyme sodium-potassium-stimulated ATP (Na+/K+ ATPase), which maintains the balance of sodium and potassium ions in living cells.

1998

- Conferences are held in China.

- James Thomson's team at the University of Wisconsin at Madison succeeds in isolating and culturing human embryonic stem cells for the first time.

1999

- Conferences held in Connecticut and Singapore.

2001

- Conferences are held in Massachusetts.

- The first nanocircuits are developed by wiring together single-molecule transistors.

2002

- GRC moves into independent headquarters in West Kingston, Rhode Island.

- Conferences are held in Maine.

- A number of research studies show that "small RNAs" control gene behavior.

2003

- Nancy Ryan Gray is appointed director after Storm retires. The Carl Storm Fellowships are established to support increasing diversity among conference participants.

- Conferences are held in Montana.

- Avastin becomes the first antiangiogenesis drug shown in large-scale clinical trials to prolong survival in cancer patients.

2004

- Researchers at Stanford and the University of California at Berkeley create the first integrated silicon circuit that successfully incorporates carbon nanotubes.

2005

- GRC produces the History and Evolution of the Gordon Research Conferences poster.

- The Huygens probe lands on Saturn's moon Titan.

2006

- 182 conferences are held at 17 sites in the United States and 6 sites overseas.

- GRC celebrates its seventy-fifth anniversary, launches the Frontiers of Science Web Site, and distributes *Reflections from the Frontiers*.

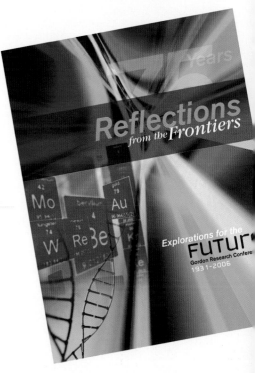

00s

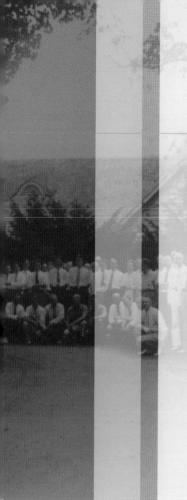

Manifest Destinations of Frontier Science

Arthur A. Daemmrich and Leah Shaper
Chemical Heritage Foundation

The Gordon Research Conferences (GRC) celebrates its seventy-fifth anniversary in 2006, providing an opportune occasion to reflect on the past and explore the future. Gordon Conferences represent a crucial but often hidden dimension of the scientific infrastructure by fostering communication across disciplinary, national, and social boundaries and by supporting pioneer research in a remarkable range of theoretical and applied fields.

GRC's beginnings in 1931 coincided with a groundbreaking period in science. New methods for physical analysis offered the first reliable quantitative approach to determining the constituents of mixtures. Chemists also were applying new physical theories of molecular bonding. Shifts in research from the organismal to the molecular level and developments in genetics signified a new phase in biology. Advances in physics laid the groundwork for the nuclear age and offered new insights into the relationship between matter and energy.

Since 1931 advances in science, technology, and medicine have fundamentally changed health care, communications, transportation, and building engineering and design, along with our understanding of environmental impact. The average human lifespan in developed nations has increased from sixty to seventy-eight years. According to the National Science Foundation, the number of Ph.D. degrees conferred in the United States in physics and astronomy, chemistry, earth and atmospheric sciences, and biological sciences grew from 859 in 1931 to more than 10,000 forecast for 2006. Thanks to the advent of the computer age, the speed of calculation has increased from one multiplication per second, performed in 1935 by the International Business Machines (IBM) 601 punched-card machine, to 135.3 teraflops (trillion computations per second), executed by IBM's Blue Gene/L supercomputer at Lawrence Livermore National Laboratory in March 2005.

Despite these remarkable changes the scientific method of determining and verifying new findings has changed little. Scientists still rely on collegial networks to discuss ideas and use both informal and formal peer-review mechanisms to validate research.

In the 1920s an intermittent set of summer chemistry meetings were held at Johns Hopkins University, where faculty members were developing new research and teaching methods that would shape science education for the nation. Beginning in 1931 graduate students and prominent academics from across the country attended more formal annual summer conferences led by chemistry department faculty and visiting specialists. Interested scientists sought permission to attend from chemistry professor Donald H. Andrews and later from Neil Elbridge Gordon, the Garvan Chair of Chemical Education.

In 1934 Gordon persuaded the university to relocate the conference series to the social club of Gibson Island, Maryland. When Gordon moved to Central College, Missouri, his former colleagues at Johns Hopkins ran the 1937 conferences in Virginia Beach. Gordon meanwhile had become secretary of the chemistry section of the American Association for the Advancement of Science (AAAS), and he promptly persuaded the organization to manage the conferences. The AAAS ran two Special Research Conferences on Chemistry in 1938 at the Gibson Island Club, and Gordon became the first official director of the conferences in 1939. By 1941, when eight conferences met, the GRC had outgrown the club facilities. Gordon identified a larger property nearby, the Symington House and Annex, and secured $8,000 from the AAAS and $33,000 from industrial sponsors for its purchase. Conferences were held on the house's majestic porch, and scientists slept in its dormitory-style rooms. In honor of Gordon's fund-raising prowess, the executive committee renamed the property the Neil E. Gordon House.

GRC's status as a participating organization of the AAAS helped ensure continuity for the Gibson Island Research Conferences (as they had become known by 1942) during a series of leadership transitions in the 1940s. Gordon became chair of the Wayne University chemistry department in 1942, and increasing involvement in university affairs led him to relinquish conference responsibilities in 1945 to his colleague Sumner B. Twiss. The following year both Twiss and Gordon resigned, and in 1947 the AAAS conference management committee, led by George Calingaert, appointed Rhode Island State College chemistry professor W. George Parks to the post of director. Once again, the conferences outgrew their meeting space, and the committee chose a new conference site, Colby Junior College in New London, New Hampshire, where ten Chemical Research Conferences were held in 1947. In the same year, Parks moved GRC headquarters to his home institution, Rhode Island State College, and recruited a young chemistry instructor and his wife, Alexander and Irene Cruickshank, to assist in operations. To commemorate Gordon's leadership, the meetings were officially named the Gordon Research Conferences in 1948.

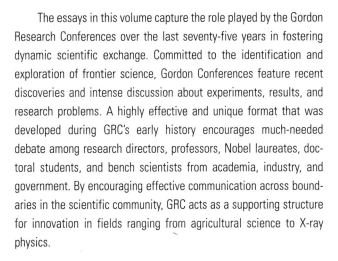

The essays in this volume capture the role played by the Gordon Research Conferences over the last seventy-five years in fostering dynamic scientific exchange. Committed to the identification and exploration of frontier science, Gordon Conferences feature recent discoveries and intense discussion about experiments, results, and research problems. A highly effective and unique format that was developed during GRC's early history encourages much-needed debate among research directors, professors, Nobel laureates, doctoral students, and bench scientists from academia, industry, and government. By encouraging effective communication across boundaries in the scientific community, GRC acts as a supporting structure for innovation in fields ranging from agricultural science to X-ray physics.

GRC's 25th anniversary year in 1956 was a milestone—four thousand scientists participated in thirty-six conferences at three New Hampshire sites, and GRC incorporated as an independent nonprofit organization. A selection and scheduling (S&S) committee was established in 1958 to advise the board on conference selection. Parks resigned in 1968, after twenty years of steady growth for the organization.

Continuing his service to GRC, Alexander Cruickshank became the third director. His forty-seven-year tenure (twenty-five spent as director) proved to be a period of extraordinary growth: both the number of conferences and the number of participants doubled. GRC continued to add new sites in New England and held its first conferences overseas in the early 1990s in Italy and Germany. Cruickshank encouraged conference chairs to apply for federal grants and other support through GRC headquarters, which enabled conferences to subsidize speakers, graduate students, and other special visitors. Further, during the inflationary pressures of the 1970s and the rapid growth of competing scientific meetings, Cruickshank kept conference fees low and maintained an informal and personal atmosphere, which appealed to scientists. He retired in 1993, and the GRC board subsequently established the Alexander M. Cruickshank lectures in his honor.

Carlyle B. Storm was selected in 1993 as the fourth GRC director, drawing on a successful career in teaching, research, and scientific management at Howard University and Los Alamos National Laboratory. International expansion begun by Cruickshank continued under Storm's leadership, with the addition of conference sites in Switzerland, England, France, the Czech Republic, Hong Kong, and Japan (see Figure 1). Storm retained the unique GRC format, promoted a "GRC brand" conference style at new sites outside the United States, and preserved the place of leading scientists in conference management. In 2002 GRC moved into its environmentally friendly headquarters in West Kingston, Rhode Island. On Storm's retirement in 2003 the board established the Carlyle Storm Fellowships to support conference attendance by underrepresented minorities and faculty from smaller colleges and universities.

Conferees often use free time to meet one-on-one or in small groups.

Nancy Ryan Gray, only the fifth director in seventy-five years, brings leadership skills from her position as director of membership at the American Chemical Society and research skills from a scientific career at Exxon. GRC continues to expand: in 2006 over 20,000 scientists will take part in one of 182 conferences at 23 sites around the world. GRC now manages a total of 350 conferences on an annual, biennial, or triennial basis (see Figure 2).

Figure 1: GRC conference sites

Hyogo (2)
Hiyama (1)
Gifu (1)
Fukuoka (2)
Hong Kong (5)
Singapore (1)

Asia

Wolfeboro, NH (230)
New London, NH (898)
Rindge, NH (4)
Antrim, NH (5)
Holderness, NH (366)
Meriden, NH (570)
Henniker, NH (90)
New Hampton, NH (513)
Plymouth, NH (429)
Andover, NH (403)
Concord, NH (2)
Tilton, NH (477)
Waterville, NH (2)

Number in parentheses indicate total conferences held at that site since 1931.

Enumclaw, WA (26)
Issaquah, WA (18)
Big Sky Resort, MT (9)
Beaver Dam, WI (13)
New London, CT (68)

Lewistown, ME (13)
Waterville, ME (20)
South Hadley, ME (38)
Williamstown, MA (7)
Bristol, RI (38)
Newport, RI (226)
Smithfield, RI (10)
Gibson Island, MD (85)
Baltimore, MD (10)

Pacific Grove, CA (1)
Los Angeles, CA (97)
Santa Barbara, CA (527)

Virginia Beach, VA (3)

United States

Prague (1)
Irsee (13)
Chantilly (1)

Oxford (85)
Roscoff (1)
Aussois (3)

Europe

Kauai, HA (2)
Oahu, HA (2)

Les Diablerets (4)
Barga (60)
San Miniato (4)
Volterra (4)

SCIENTIFIC UTOPIA BY DESIGN

Gordon Research Conferences bring small groups of conferees together for five days in secluded locations. Participants are saturated in formal scientific talks and intensive discussion sessions, balanced by shared leisure activities. Gordon chose Gibson Island for its isolated, quiet, and vacation-oriented atmosphere. On the same principle Colby Junior College was chosen for its cool temperatures, seclusion, and variety of recreational opportunities. Insulation from competing activities during conference sessions remains a key characteristic of contemporary locations. Furthermore, attendee numbers are kept small enough to promote full participation and high-quality discussion but large enough to represent a diversity of perspectives and research approaches. Limited Gibson Island facilities restricted attendance to 60 people per conference. Today attendance is usually limited to 150, unlike other scientific meetings that often host thousands of participants.

A fixed schedule of activities is key to GRC's successful operating formula. Conferences at Johns Hopkins were organized as five-week summer courses. Neil Gordon later narrowed the focus of each weeklong conference to one topic. Early conferences on Gibson Island were further structured into one or two formal lectures per day, usually held in the mornings, followed by discussion periods. By the 1940s a routine daily schedule had been established, consisting of formal presentations in the morning, discussion until lunch, and further talks and discussion periods in the evening after dinner. This schedule is still faithfully followed today with only slight modification: conferences now run from Sunday to Thursday (instead of Monday to Friday), and afternoons typically include time for poster sessions.

Integrated into this seemingly rigid schedule is ample free time. Conferees eat meals together, and afternoons and late evenings are reserved for informal discussion, relaxation, recreation, and social activities. Afternoons might be spent hiking; swimming; playing soccer, tennis, or volleyball; or white-water rafting. Jam sessions and magic shows often provide evening entertainment. Some activities have developed into traditions, like Thursday-night New England lobster dinners and softball rivalries between academics and industrialists or between scientists from the East and West coasts. By helping to cultivate personal relationships among conferees, collective leisure activities open new lines of communication.

GRC's focus on frontier science is possible because of the organization's emphasis on discussion, which is given extensive time during and after formal presentations. Such discussion provides a rare opportunity for colleagues to exchange critical face-to-face feedback. Conferees often use free time to meet one-on-one or in small groups. Debates continue over an afternoon swim or hike, during evening gatherings, and sometimes into the early morning hours.

Though set in advance, conference programs are flexible enough to address unforeseen needs for discussion. In several instances conference chairs responded to significant breakthroughs by restructuring programs or by adding extra discussion periods.

Discussions serve as a form of peer review, providing real-time feedback and dialogue at the prepublication stage, while the opportunity to make personal connections over meals or during sports activities minimizes the intimidation of constructive criticism.

A strategic mix of science and leisure activities encourages scientists to challenge each other vigorously and kindles lifelong friendships and research collaborations. Discussions serve as a form of peer review, providing real-time feedback and dialogue at the prepublication stage, while the opportunity to make personal connections over meals or during sports activities minimizes the intimidation of constructive criticism. This personal approach is particularly important for young scientists, for whom a Gordon Conference is often the first occasion to meet well-known experts, engage in serious debate, and face challenging questions in front of an audience.

Gibson Island

GRC retains its focus on cutting-edge science by encouraging and supporting new conferences in rapidly evolving fields. Applications to hold conferences in specific research areas that would benefit from the GRC format are submitted by individuals or small groups of scientists. In some instances the GRC director solicits proposals in emerging areas. New topics that do not overlap with existing conferences are first evaluated by the Selection and Scheduling Committee and then must be approved by the board of trustees. The committee also examines existing conferences using surveys of conferees, chairs' reports, and at times assessments by external monitors. Conferences whose scientific content has become mainstream—typically indicated by a lack of discussion and absence of leading-edge talks and topics—are cut to make room for new conferences.

Guaranteeing the privacy of conference communications is key to GRC's success in breaking down barriers in the scientific community. GRC's off-the-record policy, formally declared in 1937, restricts the recording or referencing of information presented during a conference, and proceedings are never published. This policy provides conferees the freedom to present novel ideas, fledgling theories, and early experimental results and receive critical feedback without worrying about being cited prematurely. Conferees are encouraged to build research on what they have learned at Gordon Conferences and to publish under their own names.

Guaranteeing the privacy of conference communications is key to GRC's success in breaking down barriers in the scientific community.

GRC's success as a catalyst for cutting-edge science is demonstrated by breakthroughs in a variety of arenas. For example, experimental techniques using electrochemical impedance spectra and current potential measurements were developed in the 1950s and 1960s at Corrosion Conferences. GRC has also contributed to new products like Nomex and Kevlar (Polymers Conference) and new medicines like protease-inhibiting drugs used to treat HIV-AIDS (Proteolytic Enzymes and Their Inhibitors). Various research collaborations, journals, scientific associations, and disciplines have also been cultivated by interactions at Gordon Conferences, such as the International Society for the Study of Xenobiotics (Drug Metabolism) and the field of electronic materials (Electronic Materials Conference). In fact, the National Science Foundation uses GRC conference topics as an indicator of current trends in scientific research.

COMMUNICATION
ACROSS BOUNDARIES

Constantly striving to be a meeting ground for all kinds of scientists, GRC has long held a nondiscrimination policy for attendance and financial assistance and actively encourages diversity in the institutional affiliations, genders, nationalities, ethnicities, scientific disciplines, and career stages of its conferees. Conference chairs are expected to attract and select a diverse group of participants, and special funding programs support the attendance of underrepresented minorities, scientists from the former Soviet Union, and faculty from primarily undergraduate academic institutions.

The organization also monitors the demographics of conference attendance to gauge its success in encouraging collaborative scientific initiatives and supporting communication across boundaries in the scientific community. These demographics provide clues to the changing dynamics between scientists in various career stages and social groups and among different types of institutions. The balance of participating academic, industrial, and government scientists has shifted noticeably during GRC's history, for example. Industrial scientists made up the bulk of conference attendance until the 1950s. An assistance fund established by the AAAS management committee in 1950, combined with the post-Sputnik expansion in federal support for government and academic institutions, shifted the balance. By 1969 there were as many scientists attending from academia as from industry. Since then academics have come to dominate GRC participation, owing partly to changes in industrial research and partly to continued expansion of university science (see Figure 3).

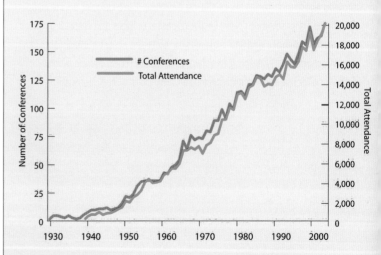

Figure 2: NUMBER OF GORDON RESEARCH CONFERENCES AND TOTAL ATTENDANCE

GRC has also witnessed an increase in the number of female scientists active in research careers. A handful of women participated in conferences in the 1940s, particularly in meetings on the biological sciences. Although staying below 5 percent until 1970, female participation increased rapidly to more than 28 percent in 2005 (see Figure 4).

Participation among scientists from outside the United States has steadily increased. Even in the face of cold war dynamics GRC strove to advance international scientific dialogue by encouraging worldwide participation, and by 1971 foreign conferees accounted for almost 14 percent of the total GRC attendance. Further encouraged by conferences located in Europe and Asia, attendance by non-U.S. scientists rose to more than 34 percent in 2005 (see Figure 4).

By creating a utopian atmosphere of scientific creativity and forging deep connections among participants, GRC catalyzes innovation in science, technology, and medicine. GRC's format fosters intensive scientific dialogue by drawing scientists from a variety of institutions, social groups, and disciplines; creating collegiality among participants; and encouraging research collaborations outside the conferences themselves. Even as the size of the scientific enterprise grows at an exponential rate, GRC builds loyalty among its participants, who return to Gordon Conferences year after year.

In its next seventy-five years GRC faces the challenges of continuing to build cohesive communities among ever-more-specialized scientific disciplines; maintaining and improving communication among scientists from industry, academia, and government; and retaining a focus on frontier science, even as the frontier expands at an accelerating rate.

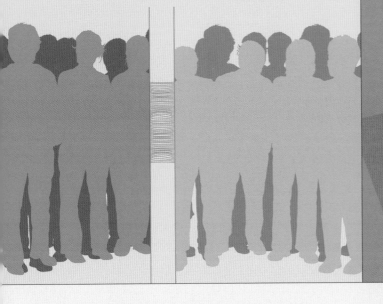

Figure 3: INSTITUTIONAL AFFILIATION OF ATTENDEES

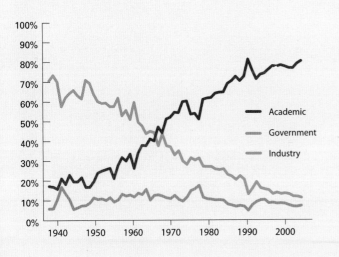

Figure 4: NON-U.S. AND FEMALE ATTENDEES

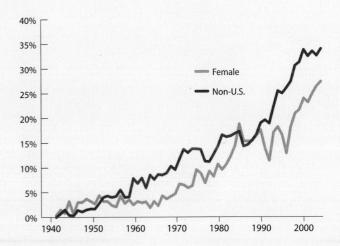

Contributor
Essays

Reflections on
Matter

Corrosion and Science

Norman Hackerman
University of Texas at Austin
and Rice University

I had the good fortune to talk with GRC founder Neil Gordon several times. He was a member of the Johns Hopkins University chemistry faculty while I was a graduate student there in the 1930s. I did not take any of his courses, but he was on my preliminary oral committee. Thus, I had a number of conversations with him in his office; in a couple of our talks he described his idea of starting what were then called the Gibson Island Conferences. He saw them as a way to move the field of chemistry forward, while providing industrial chemists with ways to reach their objectives more expeditiously.

I also had the opportunity to participate in a number of Gordon Conferences. From the late 1940s through the 1980s I attended at least one Gordon Conference each summer; in some years I attended as many as two or three. I attended the Corrosion Conference steadily. I also attended the Electrochemistry, Dynamics at Surfaces, and Chemical Education Research and Practice conferences.

The Corrosion Conferences contributed substantially to impressive uses of instrumentation along with theory and modeling to improve corrosion experimentation vastly and enhance understanding of corrosion mechanisms. The conferences also included discussions that contributed to major advances in handling corrosion problems in the field. The transfer of improved understanding of corrosion processes to field operations was particularly notable in regard to the inhibition of corrosion by the use of chemical additives.

Gordon Conference attendee interactions also brought attention to and advanced other methods of corrosion control. In fact, many of the methods currently used to explore corrosion reaction mechanisms were thoroughly discussed and advanced to their present level of understanding at the Corrosion Gordon Conferences. Early use of methods that involve spectra, microscopy, sensitive weighing capability, electrical systems insight, and purely electrochemical measurements were all discussed effectively at these conferences.

The greatest advantage of these meetings was that attendees were able to participate without worrying about being proved wrong in publication, since all discussions were officially off the record and nonattributable. Therefore, the Corrosion Conferences were effective in moving all aspects of the field forward by virtue of providing each attendee with a clear view of recent or impending developments. The same was true of the other Gordon Conferences I attended. GRC in fact had a correcting mechanism, a way to keep itself on track with only the most recent scientific progress: conferences that no longer intrigued and attracted participants were ended.

at Hand

Gordon Conferences maintain an air of informality and minimized central management that allows attendees to focus on the science at hand, a feature that many with whom I have conferred feel other scientific meetings lack. Over the years I attended as many as fifty Gordon Conferences on various concerns, and I always came away believing I had acquired something of use to my research.

From the
Cortisone Era
to the Peptide Age

It is a pleasure and a privilege to reminisce about the August 1953 Chemistry of Steroids and Related Natural Products Gordon Conference on the campus of the New Hampton School in New Hampton, New Hampshire, and the first Chemistry and Biology of Peptides Gordon Research Conference at the Miramar Hotel in Santa Barbara, California, in January 1976. These conferences differed from each other as much as New Hampton differs from Santa Barbara. What they had in common is that both were unforgettable events.

April 1948 marked the beginning of what I call the "cortisone era." That month Merck had shipped a sample of cortisone, prepared from cholic acid, to Philip Hench and Edward Kendall at the Mayo Clinic, who carried out a clinical trial to determine the effects of the two endogenous glucocorticoids—cortisone and its equilibrium partner in vivo, cortisol—on patients suffering from rheumatoid arthritis. The results showed that both hormones produced dramatic anti-inflammatory effects. These historical medical events had followed earlier chemical breakthroughs that resulted in the isolation of cortisone by Kendall and Tadeus Reichstein and the subsequent thirty-seven-step partial synthesis of the hormone by Lewis Sarett and Max Tishler at Merck.

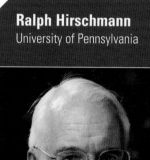

Ralph Hirschmann
University of Pennsylvania

The momentum of steroid research at the time of the 1953 Steroids Conference was the result of fierce competition among many of the world's leading bioorganic chemists who had become deeply involved in the research. Conference speakers included Derek Barton, Ian Bush, Louis Fieser, Josef Fried, Oskar Jeger, William Summer Johnson, Ewart Jones, R. Norman Jones, Durey Peterson, George Rosenkranz, Sarett, Franz Sondheimer, Norman Wendler, Robert Woodward, and other organic chemistry greats.

(continued)

From the
Cortisone Era
to the Peptide Age

Even the sense of congeniality that prevailed throughout the 1953 conference could not hide a sense of nervous tension, as each attendee wondered which of the renowned speakers might have outdone him. No scientific meeting that I attended during the subsequent fifty years could compare to that Gordon Conference. This conference was also unforgettable for me because the late Dr. Wendler, an outstanding organic chemist and my boss at the time, spoke about steroid C/D ring transformations, describing my first communication to the editor of the *Journal of the American Chemical Society* in 1952 in which I reported on the C-nor-D-homo rearrangement and coined the term *stereoelectronic control*.

Toward the end of the 1950s steroid research had peaked, but in the 1970s peptide research was greatly stimulated by growing excitement about recently reported research on peptide biology and chemistry and indirectly by news of the first total synthesis of an enzyme (ribonuclease). The latter landmark chemical event was reported in January 1969 by Robert Merrifield and Bernard Gutte of Rockefeller University and by the Merck group headed by Robert Denkewalter, my boss, and our collaborators.

William Gibbons chaired the first Gordon Conference on the Chemistry and Biology of Peptides in 1976. The Peptides Conference differed from the Steroids Conference in that the number of biologically active endogenous peptides of great importance to physiology and medicine exceeds that of the physiologically significant steroids, making peptides a very appropriate subject for Gordon Conferences. However, the intensity of the Steroids Conference in the early 1950s will remain hard to match.

Because of the intimate atmosphere that is the hallmark of all Gordon Research Conferences, taken together, the Steroids and Peptides Gordon Conferences will remain memorable for all who were privileged to attend them.

A review of current Gordon Conferences reveals a wonderful mix of the venerable and the innovative. My first GRC experience was a new conference held at Colby-Sawyer College in 1996 called Chemical Sensors and Interfacial Design (CS&ID), organized by Dick Crooks, from Texas A&M University, and Tony Ricco, then at Sandia National Laboratories. CS&ID evolved from the long-standing Electrochemistry Conference and reflected the growing field of organic-based sensing devices, including electrochemical mechanisms.

In 1996 my lab at Lawrence Berkeley National Laboratory was just coming into its own. I was honored to be invited to speak at my first Gordon Conference because I knew how prestigious GRC was. Through the CS&ID meetings I connected with and visited fellow National Lab scientists Ricco and Basil Swanson (from Los Alamos). From those interactions I realized that my work on polymeric colorimetric biosensors, in addition to its application to medical diagnostics (which had been my focus), could be applied to the important security problem of biological warfare agent detection. The off-the-record policy of GRC allowed the national need for chemical and biological sensors to come to the fore, even before the events of 11 September 2001. These discussions undoubtedly changed the direction of my research.

The Advantages of Being
Landlocked

I remember an amazing talk by Jed Harrison from the University of Alberta at the 1996 CS&ID meeting. It was the first seminar I had seen on the emerging area of microfluidics, and Jed had cool videos that went along with the great science. At the 2000 CS&ID meeting Michael Ramsey, from Oak Ridge National Laboratory, gave a talk on microfluidics that was a treat. Clearly the field was taking off. By 2001 the microfluidics community had a conference of its own—the Physics and Chemistry of Microfluidics. This is a great example of a spin-off conference that reflects the changing avenues of science.

I became a regular at the CS&ID Conference and was elected vice chair for the 2000 meeting in Ventura, California, chaired by Tom Mallouk. Nate Lewis's talk on vapor sensing using conducting polymer arrays was particularly inspiring and gave me the idea to use our colorimetric polymers in a similar vein. I could not get home fast enough! I am sure many attendees during the course of a Gordon Conference week have felt the same excitement about getting back home to pursue new ideas.

GRC is about innovative science, but it is also about people. The "free afternoons" are designed to allow casual interactions that may lead to future scientific collaborations. As vice chair of the 2000 CS&ID meeting I was charged with directing the social activities of the conferees during their free time—a task I took very seriously. I considered what to do with all the foreign scientists visiting the Golden State and came up with the perfect afternoon outing—a whale-watching trip. What better way to take advantage of our seaside location in Ventura? A boat was chartered, the conferees filed onboard, and we set sail on the turbulent waters of the Pacific. About thirty minutes into the voyage, not a single whale was sighted, but I noticed several of my colleagues turning an unusual shade of green. Soon after, eminent professors and graduate students alike were spotted recycling their lunch back to the sea. Two hours of abject misery and still no whale—not that anyone would have cared. For better or worse the "Whale Watching Trip from Hell" became part of CS&ID lore. I am convinced that communal suffering solidified the bond between attendees: many of them came back to the conference I chaired in 2001 in (mercifully landlocked) Il Ciocco, Italy.

The first CS&ID conference in 1996 cemented an emerging field of research and jump-started my own scientific career. I left academia to begin a new career in biotechnology, but I try to adopt the spirit of the GRC anytime I give a talk. Today I am honored to serve on the GRC board of trustees, and I look forward to seeing GRC lead the future of science in the next seventy-five years and beyond.

Deborah
Chiron Corp

"Cookers" in Chemistry

Fred Basolo
Northwestern University

In the late 1940s and early 1950s the field of inorganic chemistry was dominated by German chemists. My colleagues and I called them "cookers" because they described how to put new inorganic compounds together but never explained how or why the synthesis worked. At the time chemists in the United States were excited about carbon, hydrogen, oxygen, and nitrogen; in other words, they focused on organic chemistry that led to polymers and pharmaceuticals. Already in graduate school, however, I was inspired by John C. Bailar, Jr., one of only a few inorganic chemists at the University of Illinois. During World War II, American chemists working on the Manhattan Project suddenly needed advanced metals, and they soon discovered applications in instruments like X-ray diffractometers and spectrometers. American chemists who returned to their old jobs after the war (I got a teaching and research position at Northwestern University) birthed modern inorganic chemistry.

Assisted by Therald Moeller at the University of Illinois and W. Conrad Fernelius at Pennsylvania State University, Bailar had the idea to start a Gordon Research Conference on inorganic chemistry in 1950. Fernelius served as the conference's first chair in 1951. Robert W. Parry and I did most of the legwork, submitting application materials to start the conference and seeking out inorganic chemists to attend. Even though there was little funding for graduate students and faculty, in the conference's first year we attracted sixty attendees, most of whom we already knew.

Parry and I kept attending the Inorganic Chemistry Gordon Conferences because we liked the format. There were talks in the morning and evening, and we played sports or kept discussing our research in the afternoon. I often played tennis before breakfast and golf in the afternoons with Dennis Elliot from the Air Force Office of Scientific Research (AFOSR), Parry, and Parry's wife, Marj. The combination of great science and afternoons off seemed to be the reason that other Gordon Conferences continued year after year and gave us an inkling that the Inorganic Conference would also continue.

As the field of inorganic chemistry became more important, conference attendance grew quickly to a hundred. International interest also grew, and through his contacts with Denny Elliot, Parry helped arrange AFOSR funding starting in 1962 to subsidize the participation of foreign scientists. The Inorganic Chemistry GRC is still going strong, as the 180 applications to a recent conference show.

Eventually I became involved in GRC management. I was elected to the council in 1970 and then later helped review conference proposals as part of the Selection and Scheduling Committee. I became chairman of the board in 1975. Over the years Alexander M. Cruickshank and I became good friends. We always reminisced about Harold Smith, under whom we had both studied—Cruickshank at the University of Massachusetts and I at the University of Illinois.

Organized *Skepticism*

The Inorganic Chemistry GRC has always featured a variety of topics. Discussion and lectures during the early years explored crystal growth, inorganic fluorine compounds, and coordination compounds. Discussions about the lanthanides (the "rare-earth" elements) and mass-produced chemicals like ammonia, sulfuric acid, and nitric acid attracted many industrial scientists. Twenty-eight of sixty attendees in 1951, for example, were from industry and nonacademic institutions. Coordination chemistry, my specialty, eventually became a staple topic. In fact, I gave the first-ever lecture of the conference series on "Stereo-chemical Changes in Reactions of Coordination Compounds." Over the last half-decade several major fields (and their corresponding GRCs) have come out of inorganic chemistry, including solid-state chemistry, semiconductors, and bioinorganic chemistry.

The year 2001 marked the fiftieth anniversaries of both the Inorganic Chemistry GRC and my participation in it. In those fifty years I missed only six conferences. Looking back, today's chemists can see how far inorganic chemistry has come: specialties like organometallics, nanotubes, and new materials chemistry did not exist in the early days. Although there are Gordon Conferences in many of the subfields, I continue to believe in the importance of a core general Inorganic Chemistry GRC. It exposes young scientists and established specialists to new topics and different career possibilities. In sum, GRC served as a foundation for inorganic chemistry in the United States and has helped keep a core identity among its practitioners.

I went to my first Gordon Research Conference in 1951 at the New Hampton School in New Hampshire. I had just obtained my Ph.D. under John Bailar and had started working as a faculty member at the University of Michigan. Bailar did very important work in coordination chemistry and, along with Fred Basolo and Conard Fernelius, helped revive the discipline of inorganic chemistry in the United States. Fernelius was the head of the chemistry department at Pennsylvania State University; he worked with coordination compounds and organized the first Inorganic Chemistry GRC. He wanted to include young faculty at the conference, and so he invited me.

As it turned out, my first conference was a rough experience. I was fresh out of graduate school and working on a project involving boron hydrides, but I did not yet have conclusive results. I hypothesized that boron hydrides would act like coordination compounds with a hydride ion—instead of chloride or fluoride—hooked to the boron. But the compound called the "diammoniate of diborane" posed a structural problem. I thought it might be similar to coordination compounds, and I decided to present the idea without having the data to prove it. There was a lot of evidence supporting my theory, but because my presentation was still all speculation, the other researchers ate me alive. I presented an unproven hypothesis, which was not what GRC attendees expected to hear.

Robert W. Parry
University of Utah

Photo courtesy Special Collections Department, J. Willard Marriott Library, University of Utah.

Motivated by that experience, I completed my research four years later and produced evidence to show I was right. I demonstrated that one could systematically replace hydride ions with NH_3 molecules around the central boron molecule, BH_4^-. That allowed us to find group III (boron, aluminum, and gallium) compounds that were structurally and behaviorally similar to the metal coordination compounds developed by Alfred Werner in the 1890s.

For many years boron hydrides were considered anomalies because they had only three valence electrons, yet they formed such compounds as B_2H_6 and B_4H_{10}, whose formulas were identical to the carbon compounds C_2H_6 and C_4H_{10}. However, the carbon atoms of these compounds each had four valence electrons. The boron compounds were specially named "electron-deficient compounds" because they had only three conventional bonding electrons per central atom, and the multicentered

(continued)

Organized Skepticism

two-electron bond was added to chemical theory of the period to explain bonding in boron hydrides and other electron-deficient species. (The current theory also works well for "electron-deficient atoms.") Today electron-deficient compounds are an important component of modern chemistry: they form a bridging group between the metals and the more conventional two-center bonded organic groups, and many of them are excellent catalysts. In today's nomenclature these compounds contain "multicenter bonds."

Collegial skepticism was one of the most endearing features of the Gordon Conferences. If presenters had data to support their work, they could quiet the opposition; if not, they were open to questions and criticism. At the Inorganic Chemistry Conference, attendees could interrupt each other to ask a question. But the basic ground rule was that you were on your own: if you made a statement and someone wanted to question you, your answer could not be "because I say so." In all the talks I gave after my first GRC, I was very sure I had data to back up my statements.

There were many other opportunities for researchers to speak to each other at Gordon Conferences. Starting with the conferences on Gibson Island, sessions were offered in the morning and at night, leaving the afternoons free. Attendees loved having the free time: it gave rise to a lot of talk about chemistry. One of the wonderful things about the conference was the "trapped" group of excellent researchers with whom you could talk and argue. The experience is very different from an American Chemical Society meeting, where you spend ten minutes talking to presenters after they have given a talk and then one of you has to run off before you can get to know them or their science well. But if you spend an afternoon in a boat together, you discuss topics of mutual personal and scientific interest.

From ceramics and metallurgy to crystal growth and combinatorial methods, the Gordon Research Conferences have been a constant source of inspiration and community for me since 1978, when as a second-year graduate student I attended the Solid State Studies in Ceramics Conference. Over my twenty-seven years of participating in Gordon Conferences—first as a graduate student, then as a postdoc and a research scientist, and finally as a member of the GRC board of trustees—I have watched the Gordon Research Conferences influence the broader scientific community by providing what is in many cases the only venue that allows a concentrated, sustained, and open exchange of ideas.

The transformation toughening of ceramics—a process that strengthens ceramic using zirconia-based composites, heating, and cooling to change its crystal structure—was first demonstrated in 1978. It led to the use of ceramics in increasingly demanding structural applications. At my first GRC the world's leading experts on ceramic microstructures and fracture mechanics argued about what was truly known and not known about the mechanisms that interact to produce this dramatic change in mechanical behavior. I was one of six graduate students at the meeting (three men and three women), and we were all plunged into the debate. Never had I had so much fun in my, albeit short, scientific life: there were mind-stretching talks and discussions in the mornings and afternoons; broader-ranging discussions with leaders in the ceramics field at lunch, on hikes, and at the fishing hole; and the opportunity to make contact with people who, to me, had previously just been names on the pages of scientific journals.

Influencing the Course of Science

I attended either the Ceramics or Physical Metallurgy Gordon Conferences every year through graduate school and during my postdoctoral work. Each conference I attended lived up to my first—characterized by exceptional science, a collegial atmosphere, time off to enjoy the outdoors with colleagues and friends, and a chance to enter into the scientific fray in a substantive, sustained way. One of my most formative GRC experiences was seeing that I could influence the course of science even as a postdoc. During the discussion following a talk by a senior scientist, I pointed out that his (and other scientists') assertion of the inactivity of a specific mechanism had not been definitively determined by experiment. When he attempted to dismiss my comments out of hand, the discussion leader obliged him to respond explicitly. He then admitted that there was no direct evidence for the mechanism's inactivity, and he discussed what it would take to gather such evidence. During the following fall the senior scientist began a series of difficult, definitive experiments and demonstrated that the mechanism he had earlier dismissed was, in fact, the dominant one. Whether these experiments would have been done and this knowledge gained without my comments, no one can say, but it was invigorating to have had an effect on the thoughts of a scientific community.

The Ceramics Conference, which meets annually in the summer in New Hampshire, is my "home" Gordon Conference; there I have been an active participant, a speaker, a discussion leader, and a cochair (with Roger French in 1997). The conference is also home for my husband, John Blendell, who like me is a professor of materials engineering at Purdue University. He has been a GRC participant since 1976 and will chair the 2006 Ceramics Conference. GRC is a family commitment for us: I was more than eight months' pregnant, with hospital bag in hand, when we attended the Ceramics Conference in 1985. We attended the Physical Metallurgy GRC in 1986 with an eleven-month-old daughter and a supportive grandmother in tow and went again in 1987 with John's mother and our two young daughters. In the following sixteen years we worked out various strategies to keep going to Gordon Conferences sans children, including going separately in alternate years.

I have participated regularly in the Thin Film and Crystal Growth Mechanisms Conference and the Combinatorial and High Throughput Materials Science Conference and in several others occasionally, including Nanostructure Fabrication and Visualization in Science Education. All the conferences I have attended are interdisciplinary and share characteristics with other meetings. The subject matter of the Industrial Ecology GRC is an extreme example of interdisciplinary research: this conference has brought together scientists, engineers, economists, sociologists, and urban-industrial planners to tackle as a community a common set of scientific problems. As science becomes more interdisciplinary, GRC is vital in helping new fields form. A Gordon Conference in a new research area is the litmus test for determining whether conditions are right for the formation and sustenance of a new scientific community.

As a GRC board member I have had the privilege of working with scientists who represent the full range of scientific disciplines that GRC covers and are dedicated to nurturing the scientific community through first-rate Gordon Conferences. With staff members who continue to explore ways to support our community, the GRC organization is in excellent hands.

Carol Handwerker
Purdue University

Catalyzing a Discipline

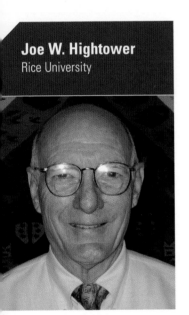

Joe W. Hightower
Rice University

When I arrived in 1959 as a Ph.D. candidate at Johns Hopkins University, I knew essentially nothing about catalysis. That fall I had the privilege of taking a stimulating graduate course on the subject taught by the late great Paul H. Emmett and knew immediately that this would be my field of research. Over the next three years I plowed through the catalysis literature, and the names of principal researchers from around the world became firmly implanted in my mind. Occasionally some of the giants in the field would pass through our laboratories at Hopkins where I might meet them, but most remained "names" to me.

Early in my final year of study Professor Emmett told me all about the Gordon Research Conferences and invited me to attend the Catalysis Conference he and Keith Hall were organizing for the summer of 1963. Without a doubt attending that conference was the highlight of my career. I met almost all the people active in my field, with many transformed from "names" into friends and colleagues. Little did I know that as a result of making those connections, I would spend a postdoctoral year with the late Charles Kemball in Edinburgh, who presented a paper at the conference, and another three years doing research with Hall, who had chaired the conference, at the Mellon Institute before launching my own career at Rice University.

I had the honor of chairing the Catalysis Conference at Colby-Sawyer College in 1973, serving as a trustee on the GRC board, and serving in 1981 as board chair during GRC's fiftieth anniversary. During twenty-five years of active service to GRC I missed only one Catalysis Conference. As a trustee I also was able to monitor a few Gordon Conferences in other areas, which introduced me to current research in fields outside catalysis. The dynamic breadth of GRC topics covers the waterfront of important fields in science.

The unique format of Gordon Conferences encourages candid presentations of novel ideas in a nonthreatening, confidential environment. Highly speculative theories frequently generate new research projects that test their validity. At the suggestion of Bob Eischens I invited R. T. K. Baker from the United Kingdom to give an exciting talk on his pioneering work with the electron microscope that showed the explosive growth of carbon filaments during catalytic reactions with small particles of transition metals. This talk resulted in Baker's move to the United States where he has continued his research for many years.

Every year I attended the conference I was fascinated by the people who occupied the front rows in the lecture hall and never failed to present challenging questions to the speakers. At the risk of omitting many important researchers, these included Emmett, Hall, Herman Pines, Bob Burwell, Val Haensel, Sol Weller, Sam Siegel, John Sinfelt, John Peri, Dick Kokes, Wolfgang Sachtler, Charlie Plank, Frank Ciapetta, Pierce Selwood, Bert Farkas, Werner Haag, Alex Mills, Hans Benesi, Jack Lunsford, and several others, all of whom have served as chair of the Catalysis Conference.

Organic Reactions
Beget Scientific Interactions

I will never forget the social benefits of the Gordon Conferences. Technical presentations were held in the morning and evening, while "free afternoons" provided time for one-on-one discussions about research and extracurricular activities. Some of my favorite experiences, when not talking about catalysis, were jogging, playing tennis, swimming in Lake Sunapee, climbing Mt. Kearsarge, and enjoying an off-campus lunch with Alex Cruickshank. It was at the Catalysis Conferences that my wife met many of my colleagues, and, of course, the lobster dinners on Thursday evenings were a gourmet's delight. One of my social contributions was to write words to favorite classic songs for a sing-along after dinner. I was pleasantly surprised to learn that many of my Catalysis Conference colleagues enjoyed this activity.

Through discussions at GRC other national and international catalysis organizations were born, including the International Congress on Catalysis, the North American Catalysis Society, and many local catalysis clubs like the one I belong to in the Southwest. Many of these organizations would never have materialized without the platform for discussion provided by the Gordon Research Conferences. Long live the GRC!

As a graduate student at the University of Chicago in the late 1950s, I had heard glowing accounts of the Gordon Research Conferences from several of my professors. So with great anticipation I applied to attend my first Gordon Conference, called Organic Reactions and Processes, in 1960, six months after arriving at Bell Laboratories in Murray Hill, New Jersey. Although I knew this conference was very popular, I was disappointed to hear that I had been put on the waiting list. There was a GRC rule at the time that a founding sponsor of the organization had the prerogative to send one representative to each conference. Bell Labs was such a sponsor, and since no one else from there had applied, I became the organization's representative at Organic Reactions and Processes. Immediately I became aware of the unique features of Gordon Conferences that stimulate informal discussion and encourage what we today call "networking." These same features are what bring me back to the conferences year after year.

Anthony M. Trozzolo
University of Notre Dame

There I met and heard many researchers in the field I had known about only through their publications. It was a great experience for someone just out of graduate school. Among the speakers was Phil Skell of Pennsylvania State University, who spoke about the differences in reactivity between the singlet and triplet states of diphenylmethylene. Shortly after my return from the conference, Bob Murray (now at the University of Missouri in St. Louis) and I discussed at a Bell Labs blackboard how interesting it would be to make a molecule with two diazo groups and what multiplicity one might expect from the dicarbenes generated. We synthesized a number of novel diazo and bisdiazo compounds and, in collaboration with Ed Wasserman and Bill Yager, characterized the structure and multiplicity of the wide spectrum of carbenes formed by decomposition of the diazo compounds.

At the 1960 conference I met Bill Mosher, who was elected to chair the 1962 conference. I was elated to be invited by Bill to present our results on the novel carbene and dicarbene work at the 1962 conference. I am certain that this occasion played an important role in increasing the visibility of our work, and subsequent Gordon Conferences provided a vehicle for presenting further developments in these studies.

(continued)

Organic Reactions
Beget Scientific Interactions

GORDON RESEARCH CONFERENCE
New Hampton School
ANALYTICAL CHEMISTRY
I.M. Warner, chairperson
L.D. Rothman, vice chairperson
August 3 - 7, 1992
Achber Studio, Laconia, N.H.

In 1962 GRC director W. George Parks published a notice in *Science* soliciting proposals for new Gordon Conferences. I sent in a proposal citing the fact that no conference was being held on a continuing basis on organic photochemistry and that the subject was developing rapidly. A chance meeting at the luncheon of the 1963 Metro Regional Meeting in Newark with Cecil L. Brown, then on the GRC board of trustees, helped to reinforce the case. The Organic Photochemistry Gordon Conference was approved to be held for the first time in 1964 at the Tilton School in Tilton, New Hampshire. The meeting's first speaker was the late Sir George Porter, who was awarded the Nobel Prize in chemistry three years later. The conference prospered, with many of the field's developments receiving early exposure there. The conference has since become international in scope and participation.

Organizing the first meeting of a Gordon Conference series privileges the organizer to choose speakers from the entire field, since none have spoken there before. This opportunity has resulted in my lifelong friendships with speakers and conferees alike. Interestingly, my chairman's award certificate was signed by Cecil L. Brown; ten years later I gave the Cecil L. Brown Lectures at Rutgers University.

During my tenure at Bell Labs I was occasionally asked by Bill Slichter, the Bell Labs representative on the GRC council, to attend council meetings in his absence. There I grew to understand the personal qualities of Alex Cruickshank that made the Gordon Research Conferences so successful under his directorship. After I moved to Notre Dame in 1975, I was elected to the council and later to the Selection and Scheduling Committee, and in 1988 I became a trustee. Those assignments provided the opportunity to interact with a large segment of the chemical community. Although it is difficult to establish firmly a cause-and-effect relationship, I have no doubt that those exposures contributed in a number of ways to several milestones in my career, including becoming the Huisking Chair at Notre Dame and serving as editor of *Chemical Reviews*.

As a board member and during my fourteen-year stint on the Selection and Scheduling Committee, a number of significant changes took place at GRC. Poster sessions were added to increase conferee discussion and participation. GRC went international, with conferences meeting first in Europe, then in Asia. I also had the opportunity to meet a dedicated group of board members, who along with Alex Cruickshank, Carl Storm, Nancy Gray, and the GRC staff, have contributed so much to the history and development of the Gordon Research Conferences. The conferences now enjoy international respect and serve as a model for many other scientific meetings. However, the deepest impression left on me from these interactions is the family-like atmosphere I have had the pleasure to be a part of for the last forty-five years. Since Neil Gordon first used the special conference format that included informal discussion, the conferences have had a vital impact, not only on the careers of individuals but also on the frontiers of science of every discipline represented by GRC.

Newly minted assistant professors of chemistry and new analytical chemists in government and industry came to the Analytical Chemistry Gordon Research Conference in a variety of ways. Isiah Warner was first introduced to the conference in 1978 as an assistant professor of chemistry at Texas A&M University. One of his colleagues, Thomas Vickery, wanted to know if he was going to attend the upcoming Analytical Chemistry GRC. Warner was vaguely familiar with the conference, but not in any great detail. Vickery quickly filled him in on the essentials. Warner soon attended his first GRC, beginning a love affair that would last for decades and culminating with his chairing of the Analytical Chemistry GRC in 1992.

The field of analytical chemistry was undergoing a sea change from sulfide qualitative schemes, volumetric and gravimetric analysis, and electrochemistry. Analytical chemistry grew to include various methods of spectroscopy and chromatography, such as nuclear magnetic resonance; infrared and atomic absorption; ultraviolet absorption and fluorescence; X-ray absorption, fluorescence, and diffraction; and plasma emission. Other techniques include mass spectrometry, two-dimensional high-performance liquid chromatography, liquid chromatography–mass spectrometry, gas chromatography, and liquid chromatography. Each method became a new subgroup with its own experts to interpret its results. Communication across these groups became very difficult, and so Mary Wirth changed the conference outline in 1996 to reflect a problem-oriented rather than a topic-oriented approach.

Humanizing SCIENCE

Isiah M. Warner
Louisiana State University

James W. Robinson
Louisiana State University

Mary Wirth
University of Arizona

Analytical chemistry is a subdiscipline of the broader field of chemistry. Its goal is to answer the question, "What is this and how much is present in the sample?" Analytical chemists go to analytical meetings and publish in analytical and other chemistry-related journals. Most lab technicians only make analytical measurements, but the researchers develop new fields; that is, they develop new techniques, new instruments, and new applications to solve new problems at ever-decreasing concentrations and in the most nondestructive way possible. All analytical researchers must be able to cite the reliability of their data in rigorous detail, while always remembering that discussion of observations alone is philosophy. When hard data are added, it becomes science, and analytical chemistry provides the bridge between the two.

From the beginning Analytical Chemistry Gordon Conferences were designed to bring together representatives from industry, government, and academia, to ensure cross-fertilization between these sectors. Every two years a chair from academia is elected, and a chair from government or industry is elected in the alternating years. This helps ensure a good mix of analytical chemists and fields. The theme chosen for each conference is usually a cross between the chair's area of expertise and the current "hot" analytical fields, including nanotechnology; medicinal, pharmaceutical, and forensic chemistry; and DNA analysis. Here experts can meet the leaders of their fields and discuss the problems of their particular topics.

The most rewarding experiences by far for all of us have been hearing outstanding scientists present their research from first principles. They made it logical and easy to understand—an ability that is the mark of true genius. One of the most memorable speakers was Milos Novotny, who first described capillary high-performance liquid chromatography and its use in distinguishing diabetics who do not have discernible medical symptoms from nondiabetics. (Strangely enough, medics objected to this technique on the grounds of it being "empirical.") Other exciting speakers were Alex Pine, who spearheaded solid-state nuclear magnetic resonance; Egil Jellum, who introduced isodalt chromatography for cancer diagnoses; Dick Zare, who discussed the interaction of light with matter; Walt McCrone, who gave insight into the Shroud of Turin; Lynn Jelinski, who described how nerves transmit signals; Jim Callis, who first used near infrared to assess skin burns; John Fenn, who discussed his now Nobel-winning technique of electrospray ionization in mass spectrometry; and Alan Marshal and Graham Cook, who spoke on ion trap MS. The list of field leaders goes on and on.

Often the speakers never finished their presentations because there was so much discussion. Many friendships were formed with people from around the world. On Thursday nights we had a fine buffet dinner (in the early years you could have all the lobster you wanted). Finally, during the evening session, which was devoted to general science, the election took place. This event was followed by the "turtle test" for all new attendees, which opened with the question, "Are you a turtle?" Rather than acting as an extension of a textbook, the Analytical Chemistry Gordon Conference succeeded in personalizing and humanizing science. In the end the motto of the conference was always "discussion, not questions and answers."

GC: Gordon Conference or Gas Chromatography?

Fred W. McLafferty
Cornell University

I started attending Gordon Research Conferences in the early 1950s—mainly the Analytical Chemistry Conferences. I was then working on mass spectrometry in the Dow Chemical Company Spectroscopy Laboratory in Midland, Michigan. The Analytical Chemistry Conference in 1954 was particularly memorable, as it was there that I heard about gas chromatography (GC) for the first time. The talks on GC by Steve Dal Nogare of DuPont and H. N. Wilson of ICI (Imperial Chemical Industries) were so exciting that I could hardly think of anything else. I spent much of the remaining free time asking them questions. I was even lucky enough to take the train back to Boston with them, filling a notebook on how to build such an instrument, how to pack columns, where to buy a detector, and so forth. Although I lost track of Wilson, Steve was always willing to provide solutions to problems and was a very good friend until his tragic death.

As soon as I got back to the laboratory in Midland, Roland Gohlke took over the gas chromatograph construction project with high enthusiasm and unusual creative energy. He soon not only had a working instrument, but he also found so many immediate applications at Dow in both research and production that he began "manufacturing" gas chromatographs. His innovations included putting the instruments inside a one-gallon Dewar vessel for temperature control, packing columns with Tide detergent, and persuading the Gow-Mac Instrument Company to make a special detector. His product was so successful that he placed two hundred of his instruments at Dow before instrument companies began competing in gas chromatograph production.

Gohlke's gas chromatograph was also the first to be directly coupled to a mass spectrometer (MS) that could measure the mass spectrum of an eluted fraction in real time. In February 1956 Roland and I drove a gas chromatograph and samples from Midland down to the Bendix Research Laboratories in northern Detroit where William C. Wiley and Ian H. McLaren had developed the first commercial time-of-flight (TOF) MS. As described by Wiley, Dan Harrington, Roland, and myself in the *Journal of the American Society for Mass Spectrometry* in 1993, the TOF MS measured mass spectra at a 10-kilohertz repetition rate, displayed on an oscilloscope and recorded by a Polaroid camera. For the first GC-MS run, however, Roland called out each compound immediately as its mass spectrum was displayed, a technique he described in *Analytical Chemistry* in 1959.

Although corporate secrecy in analytical techniques is not as important as it is in product manufacturing, DuPont and Dow were broad-minded in encouraging the open communication that made the Analytical Chemistry Gordon Conferences highly valuable. When I asked Ray Boundy, then Dow's director of research, about this he laughed and said, "After all, DuPont is our best customer." Improving communication was also a major objective of Dow's in establishing their Eastern Research Laboratory outside of Boston, and in June 1956 I was sent there as its first director to set it up. Of course the laboratory's contribution to Dow research was directly related to its success in keeping up with the best in science. An important long-term benefit to Dow was also the hiring and development of scientists with excellent skills in research and communication.

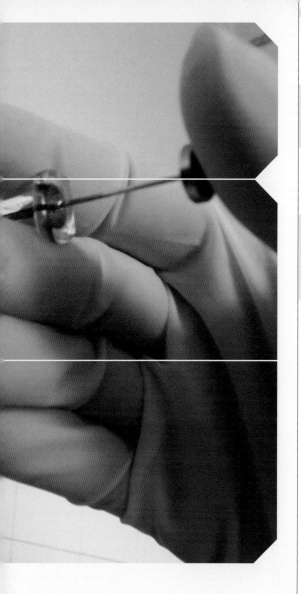

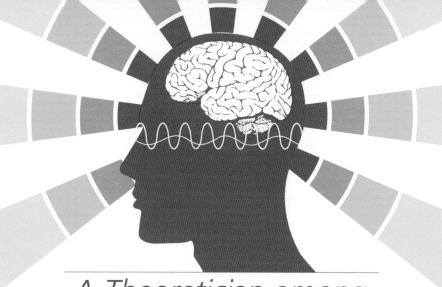

A Theoretician among Experimentalists

It is critically important to keep reexamining the efficiency of the scientific enterprise, finding ways to get more results from research and development in various fields, and better communication is a key factor in this objective. Complaints about inadequate communication are often heard in the scientific community, such as "meetings are too large," "it is hard to talk to people," "we didn't become well enough acquainted to really discuss things," and "if only we had more time." Available for more and more fields every year, the Gordon Research Conferences are the solution.

Early in my career I started going to scientific meetings and found that they opened new avenues for my research. I was fortunate to have completed two major theories, RRKM (Rice-Ramsperger-Kassel-Marcus) and electron transfer, by the time I was thirty-three. In fact, I attended my first two scientific meetings in order to present the RRKM (in 1950) and electron transfer (in 1956) theories. During those years I was a postdoc and, after 1951, a young professor at the Polytechnic Institute of New York. I had completed my B.S. and Ph.D. in chemistry at McGill University and had spent a few years working on the RDX Project (an investigation into cyclotrimethylene trinitramine, a common army explosive) and at the National Research Council of Canada. I did my postdoctoral studies in theoretical chemistry at the University of North Carolina (UNC), where my explorations into theory were very exciting and rewarding, owing to my special previous interest in mathematics.

Rudolph A. Marcus
California Institute of Technology

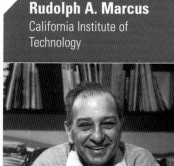

While at UNC, I worked off the classic RRK (Rice-Ramsperger-Kassel) theory of the 1920s and combined it with transition state theory of the mid-1930s. This synthesis later became known as RRKM theory, which provided a way to determine the rate coefficient of a unimolecular reaction. It was also great to take the electron transfer theory to the first Electrochemistry GRC in 1964 because most of the people there were experimentalists working on a wide variety of electrochemistry topics, ranging from hydrogen evolution to semiconductors.

(continued)

A Theoretician among Experimentalists

I attended various Gordon Conferences because my theoretical work overlapped many fields. At each conference I interacted with different scientific communities. For example, I did some work in electrochemistry examining electron transfer reactions, both in solution and in biological systems. I also worked on gas-phase unimolecular reactions. Each of these fields involves completely different groups of people. In fact, when it became known that I had received the 1992 Nobel Prize in chemistry, some people did not know whether I had received it for RRKM theory or for electron transfer theory.

In the mid-1950s, when I was teaching at Polytechnic, I started to do the work on electron transfer theory for which I was later awarded the Nobel Prize. I was able to help determine the rate of chemistry's simplest chemical reaction—the transfer of one electron. Electrons are transferred during oxidation and reduction reactions. One of my contributions was to show how much reorganization energy—the energy used for the reacting molecules and solvent to reach structures that allow the electron transfer to occur—is required in one of these simple reactions. I was also able to determine the activation energy, the energy needed for such a reaction, and many properties relating the rates of different reactions. Electron transfer theory was the basis for much of my subsequent work.

Gordon Conferences not only made it possible for me to have contact with experimentalists, but they also showed experimentalists a way to interpret and understand more of the results of their experiments, and even predict new ones. Most important, Gordon Conferences bring together people who are working on related fields and provide a forum for discussion. GRC's main principle is to promote the exchange of ideas, and its goal is to drive participants to learn how to explain their ideas to others. By having to explain my work to experimentalists at GRCs, I was really forced to think things through.

> Reflections

"Accelerator mass spectrometry (AMS) was first brought to the attention of the drug metabolism field at the 1997 [Drug Metabolism] Gordon Conference. Since then, AMS has been steadily growing as a way to approach clinical studies of drug metabolism without the use of radioactive doses. At least one company has been formed to provide the service, and so far several pharmaceutical companies have used it in drug development."

Ronald White
Schering-Plough Research Institute

The Ultimate SOUNDING BOARD

My colleagues and I published our first paper on isotope effects in 1984. By that time physical organic chemistry made it clear that, in principle, transition-state structures for enzymes could be obtained. We had been inspired by a paper by the biochemist Richard Schowen on the catechol O-methyl transferase enzyme: he had used a mapping technique based on kinetic isotope effects to locate the enzyme's transition state.

Frank Mentch was doing the computational chemistry in my lab for our first effort to obtain the transition-state structure of another enzyme, AMP nucleosidase, and we flew in a small airplane to visit Richard Schowen to get his critique of our work. Schowen invited me to speak at the Chemistry and Physics of Isotopes Conference, and I have attended every Isotopes Conference since. The conference has been the sounding board for our research on enzymatic transition-state analysis. Some of what we were doing was quite speculative, and the expertise we have been exposed to at GRC has been invaluable in critiquing and validating our approaches to enzymatic transition states.

We were soon confident that robust quantum mechanical models of enzymatic transition states were readily accessible by these methods. The logical extension of this knowledge was to design and synthesize stable molecules of the transition states. This was also surprisingly successful and yielded molecules that are powerful inhibitors and mimic the transition states.

Most recently we have been applying these principles in a logical way to design and develop new agents for human disease. The process involves selecting an enzymatic target implicated in human disease, measuring its intrinsic KIE (kinetic isotope effect), and solving its transition-state structure; using the transition-state structure to design chemically stable mimics of the transition state; synthesizing the transition-state mimics; and testing the mimics in enzymatic and biological systems. This approach has yielded a transition-state inhibitor now being tested against T-cell cancers in FDA-approved clinical trials at thirty-seven clinical sites in the Americas and Europe. A second transition-state analogue to defend against tissue transplant rejection and autoimmune diseases is also in preclinical trials. A third is in animal trials to test its effectiveness against prostate and head and neck cancers. Others are in development as agents against antibiotic-resistant bacteria.

The Isotopes Conference has been an important venue in which to preview the fundamental science of these developments. Today it is clear that KIE, enzymatic transition states, and the directed design of transition-state analogues will bear significantly in the treatment of human diseases. The conference has been an important factor in nurturing and developing this fundamental science. It is quite unexpected and surprising that in less than twenty years our basic interest in kinetic isotope effect analysis for enzymatic mechanisms has led to new agents against human diseases that are now in clinical trials.

Vern Schramm
Albert Einstein College
of Medicine

WHAT? INORGANIC LIFE?

Harry B. Gray
California Institute of Technology

I was born in a southern Kentucky tobacco farming community in the middle of the Depression. Encouraged by my mother's two brothers, who dabbled with ham radios on the farm, I took up chemistry at the age of twelve. With big bottles of sulfuric and nitric acids from a supply house in Chicago, as well as many other chemicals, I could claim that I had a fairly decent lab in the basement of my house. Of course I made my mother very nervous by running chemical reactions that often gave dramatic results. It was around that time that my very dear grandmother lost her life to cancer. She lived on the farm, cooked for me, and did everything for me, so I was very close to her. It was a defining moment in my life, as I knew then that I wanted to become a scientist. It was apparent to me that there were diseases that could take loved ones away too soon, and I thought that by becoming a scientist I might be able to do something about it.

After college in Kentucky I went to graduate school at Northwestern University, where I worked with Fred Basolo and Ralph Pearson on platinum reaction chemistry. During that time I became very interested in crystal field theory as it helped me to understand the colors and magnetic properties of inorganic compounds. It was an exciting time in inorganic chemistry. After I finished my degree, I accepted a National Science Foundation postdoctoral fellowship to work on crystal field theory with Carl Ballhausen in Copenhagen. Soon after I arrived, however, I turned to molecular orbital theory as a more powerful framework for my investigations of the electronic spectra of inorganic complexes. I made good progress in my work on the electronic structures of metal-oxo complexes, and after an extensive discussion with Martin Karplus, who was a visitor in Copenhagen, I was invited to join the

chemistry faculty at Columbia University. I started my academic career in the Morningside Heights neighborhood of New York City in the fall of 1961.

I knew from earlier conversations with Fred Basolo that the Inorganic Chemistry Gordon Conference was very special. He urged me to go as soon as I had a chance. My very first Gordon Conference was Inorganic Chemistry in the summer of 1962 at the New Hampton School in New Hampton, New Hampshire. I had been invited to give a talk on my work on inorganic electronic structure by Bob Parry, who was organizing the conference that year. Bob also invited Dick Holm, Dick Carlin, Al Cotton, and Dick Fenske—outstanding investigators who were doing pioneering work in inorganic chemistry. This mix made for an incredible meeting; there were many heated discussions because we all had different ideas about the best theoretical model to use to explain the ground and excited states of inorganic complexes. Was molecular orbital theory better than crystal field theory? Or should we use the best parts of both, which we called ligand field theory? And with molecular orbital theory what level of theory was useful for inorganic complexes? With the field in flux we were looking for some common ground in a theory that would fit experimental results for many classes of inorganic compounds.

In addition to the sessions and intense scientific discussions there was much time for tennis. The tennis courts at the school were great. One year Gil Haight and I even entered the New Hampshire Open and did quite well. We were semifinalists in doubles and could have gone farther, but we managed to blow a crucial set by talking too much to each other. My experiences at the Inorganic Chemistry Gordon Conference really hooked me.

I started going to Gordon Conferences every year, with Metals in Biology being one of my favorites. In more recent times I have also been a regular at Inorganic Reaction Mechanisms and Quinone and Redox Active Amino Acid Cofactors (now known as Protein Derived Cofactors, Radicals, and Quinones). Of course I like to go to the "original" conference (Inorganic Chemistry) whenever I can. I have also made more than one appearance at Organometallic Chemistry, Electron Donor Acceptor Interactions, and Electrochemistry. Indeed, I have been told by Carlyle Storm that I have the all-time record for attendance at Gordon Conferences.

The Gordon Conference on Metals in Biology was instrumental in building a bridge across the divide between inorganic chemistry and biology. Paul Saltman was the first chair to welcome "inorganikers" in large numbers to the conference. At Paul's meeting, where I was asked to talk about our work on iron in biochemistry, I remember many fruitful discussions with Jack Halpern on problems that we felt could be attacked by inorganic chemists entering the field. Many of my own contributions to bioinorganic chemistry have come from experimental work that has shed light on the underlying physics and chemistry of electron transfer reactions that occur in photosynthesis, respiration, DNA synthesis, and many other biological redox processes. Much of this work was presented for the first time at Gordon Conferences. The discussions I had at those conferences often changed the direction of my work—always for the better, I am sure.

Bioinorganic chemistry is now booming, thanks in large measure to the hard-hitting discussions that have taken place between biochemists and inorganic chemists at Gordon Conferences. The field is a vibrant, ever-changing area of science that includes metals in medicine, of which the platinum-based cancer drugs that are famous for their role in restoring the cyclist Lance Armstrong to full health are a great example. I had worked on similar platinum (II) complexes in grad school, so I was particularly pleased when I learned that the cis-dichloro-diammine complex was being used very effectively in the fight against the most dreaded of diseases. All in all, bioinorganic chemists have made many important contributions to science and medicine that have come about because of a deeper understanding of the structures, spectroscopic properties, and reactivities of metal-containing biomolecules.

The Gordon Conferences have played a very special role in science. It is a pleasure to have an opportunity to thank the wonderful folks who have made them the best conferences on the planet.

STARTING OUT ON THE FRINGE

Joan Selverstone Valentine
University of California at Los Angeles

I have attended Gordon Research Conferences for so long that I cannot recall for certain which one I attended first. I believe I started out at the Inorganic Chemistry and Organometallic Chemistry Gordon Conference. I remember that there were many men and very few women at the early conferences. (I recall being one of only two women attending one conference, and our beds were in the air-conditioned infirmary instead of in the hot and humid dorms—who could complain about that?) More memorable, however, were the excitement and the intensity of the discussions about science at lunch, in the afternoon, and late into the night. As a young scientist I was enormously inspired, even if by the end of the week thoroughly exhausted, by these experiences.

Oddly, it was a Metals in Biology Conference that I did not attend that had the biggest impact on my research career. As a student I loved inorganic coordination chemistry, particularly the compounds with beautiful colors. I had no interest in biochemistry then; moreover, bioinorganic chemistry was unknown to me at the time. In the early 1970s, when I was an assistant professor at Rutgers University, I was looking for something new to work on, and so I called on my friend Tom Spiro for advice. He had just returned from the Metals in Biology Gordon Research Conference, where biochemist Irwin Fridovich had reported a strange new copper-containing enzyme he called superoxide dismutase. Knowing that I had studied metal dioxygen complexes, Tom suggested that I study the reactions of superoxide with metal ions and complexes. That conversation set the course of my research for at least the next thirty years. As my interests became more biologically oriented, I started to attend the Metals in Biol-

ogy GRC, which rapidly became my primary conference. It was the birthplace of bioinorganic chemistry (short for biological inorganic chemistry), a field now so big and so well-established that it is difficult to believe it was "on the fringe" in its early years and even antedated the popularity of interdisciplinary research.

Considered neither fish nor fowl, we bioinorganic chemists started out making model compounds, but soon we branched out into the study of biological molecules themselves, despite the fact that most of us had little or no biochemical training. In those days the only conference where I really felt at home was Metals in Biology, not only because of our common research interests but also because of the substantial number of women who participated in this meeting from its inception.

Recently I discovered that going to a Gordon Conference in a field I have never explored is a great way to become knowledgeable in that area. It is like learning a new language by total immersion. Further, when I cannot understand what anyone is talking about at the dinner table and the questions I ask stunningly display my ignorance, my anonymity at a conference where no one knows who I am is a distinct advantage. Nonetheless, I know of no better way to learn the background of and the way people think about a field—reading papers just is not the same. Among the particularly memorable meetings of this sort were Protein Folding Dynamics, CAG Triplet Repeat Disorders, and Cell Death.

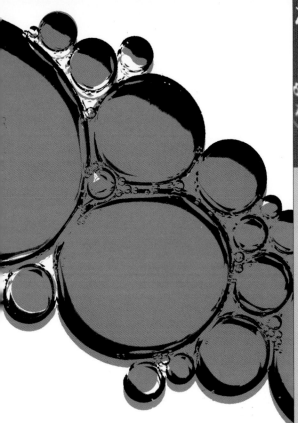

Metals in
BIOLOGY
and its progeny

In the early 1970s, as an assistant professor at the State University of New York at Stony Brook, I started attending both the Inorganic Chemistry Gordon Conference and the Metals in Biology (MIB) Conference. I had only a faint awareness of the diverse and critical roles that metal ions play in biological systems. Attending the MIB Conference convinced me of the area's enormous potential, and I committed myself to the nascent field of bioinorganic chemistry.

Edward I. Stiefel
Princeton University

Three specific results came from my early attendance at MIB. First I became aware of the ignorance about critically important molybdenum enzymes, such as nitrogenase, nitrate reductase, and xanthine oxidase, and decided to focus on this area of research. Second, on the request of Steve Lippard (then at Columbia University and editor of *Progress in Inorganic Chemistry*) with whom I drove to the conference, I wrote a review called "The Bioinorganic and Coordination Chemistry of Molybdenum" that became a citation classic with over eight hundred citations. Third, an invitation to attend a seminar from Bill Newton of the Charles F. Research Kettering Laboratory, who I met on the tennis court during the MIB Conference, led to my joining the Nitrogen Fixation Mission at Kettering. Clearly, MIB had a tremendous influence on my early career and research.

In 1980, while still attending MIB, I moved to Exxon Corporate Research, and I persuaded the company to provide some support for MIB for many years. In 1993 I chaired MIB, and, as Doug Rees presented the first nitrogenase crystal structure, I realized how far the field had come but how much was still unknown. MIB continued to meet on a yearly basis in California, but there were always good-natured complaints that only one session, at most, was spent on a particular subfield. This opened the door to new Gordon Conferences, such as the Nitrogen Fixation Conference, chaired by Doug Rees and Bill Orme-Johnson, that first met in 1994 and at which I gave the opening talk, called "Chemistry of Nitrogen Fixation." It was great to have a Gordon Conference focused on the molecular aspects of nitrogenase.

(continued)

Like many other scientists, I have a primary, or home, GRC, and for me this remains Metals in Biology. I attend whenever I can to learn the latest, to discuss new ideas with people I trust and respect, to meet the newcomers to the field, and to catch up with friends. Yet there are other Gordon Conferences that have proved essential to my research because of what I learned and who I met. Those meetings include Metals in Medicine, Oxidative Stress and Disease, Oxygen Radicals, Protein Derived Cofactors, Radicals and Quinones, and Enzymes, Coenzymes and Metabolic Pathways.

Each year I look forward to reading the schedule of upcoming Gordon Conferences to find out who is speaking and what the hot new topics are. For me there is no better indicator of where science is today and what its future might be.

METALS IN BIOLOGY AND ITS PROGENY

At the Nitrogen Fixation Conference I first realized that a GRC on other molybdenum enzymes would attract a large audience. A few years later Russ Hille, of Ohio State University, and I proposed a new Gordon Conference called Molybdenum and Tungsten Enzymes. Together we chaired the inaugural conference held in New Hampshire in 1999, with eighty attendees. Two years later one hundred attendees met at Oxford University. This specialized subject attracted bioinorganic chemists, crystallographers, biophysical chemists, molecular biologists, microbiologists, botanists, and physicians. The resulting cross-fertilization was remarkable, and the conference continues to meet on a two-year cycle.

In 1998 my group at Exxon joined researchers from Princeton University, Rutgers University, and the University of California campuses to form the Center for Environmental Bioinorganic Chemistry (CEBIC), directed by François Morel and funded by the National Science Foundation and the Department of Energy. Many interactions among CEBIC researchers were first forged at Gordon Conferences. CEBIC was such a wonderful experience that in 2001 I moved to Princeton University with appointments in the chemistry department and the Princeton Environmental Institute (which houses CEBIC). At Princeton I have been fortunate to teach a freshman seminar called Elements of Life, which has convinced me that bioinorganic chemistry is a splendid vehicle for teaching chemistry and its relation to biology, geology, astrobiology, and environmental science. Meanwhile, CEBIC continued to thrive, and each year our summer workshop at Princeton drew more people from outside CEBIC. This success led François Morel and me to propose the Environmental Bioinorganic Chemistry (EBIC) Gordon Conference, and in 2002 François and I chaired its inaugural conference, with 125 attendees from a remarkable range of disciplines. The conference now meets regularly on a two-year cycle.

Meanwhile MIB is thriving, despite the heavy attendance draw (one thousand participants) of the International Conference on Biological Inorganic Chemistry. Without losing momentum, MIB continues to spawn new conferences such as the 2002 Metals in Medicine (chaired by Nick Farrell) and the 2005 Cell Biology of Metals (chaired by Nigel Robinson and Dennis Winge). MIB is far from finished as it continues to stimulate interdisciplinary cross-fertilization of ideas in an intimate setting with a wonderfully paced program that only GRC can provide.

> Reflections

" GRC first started me thinking about a "deep biosphere" with respect to methane generation in sediments. Our group successfully carried out initial experiments demonstrating in situ organisms in deep sediments. GRC helped me make interdisciplinary contacts that allowed us to do this work. Later, as a result of conversations between myself and James Maxwell, John Parkes in James's lab in the United Kingdom was encouraged to continue and extend these experiments. Deep biosphere science is currently a very hot research area thanks to John's pioneering work! "

Jean Whelan
Woods Hole Oceanographic Institution

A *Lively* FIELD FROM *Lively* MEETINGS

Harry Gray

The field of bioinorganic chemistry and the Metals in Biology (MIB) Gordon Conference have grown hand-in-hand since the 1960s, when biologists and biochemists like Paul Saltman and Bert Vallee began to realize the importance of inorganic compounds in their work. Paul Saltman was the principal mover and shaker who made the MIB Gordon Conference, which began in 1962, the very successful gathering of the faithful that we know and love today. What an impression Paul made! I had the misfortune of being scheduled to speak after Paul at one of the early MIBs; Paul had the audience so in tears from laughing that my talk was a loser from the start. Chemist Jack Halpern yelled from the front row, "Harry, you're no match for Paul!" Indeed I wasn't, but bioinorganic chemistry was moving so fast I didn't worry about it!

In the 1960s and 1970s MIB sometimes became a battlefield for debate between inorganic chemists and chemical biologists. We had to rotate the chairs, one meeting chaired by a biochemist, the next by an inorganic chemist. The field of biological inorganic chemistry, however, would not have coalesced in the same way if it were not for the Gordon Conferences. Leading scientists would never have gotten together and duked it out in another setting. At the Gordon Conferences they were confined for a week and forced to talk to each other, so it couldn't not happen.

By the mid-1990s MIB had become very popular and far oversubscribed. Nearly all the participants were established investigators; few graduate students or postdocs applied to attend, as there was little chance they would be accepted. Members of the GRC board, while pleased with the overall success of MIB, were concerned about the lack of student participation. Carl Storm, then the GRC director, took the initiative and proposed that GRC hold a Bioinorganic Chemistry Graduate Research Seminar (GRS).

On Carl's request Joan Valentine (who was on the Selection and Scheduling Committee at that time) recruited Edward H. Ha, a student from her group at the University of California at Los Angeles (UCLA), to organize the first GRS meeting in 1996. Ha and Joy Goto, also of UCLA, cochaired the meeting, assisted by Gretchen Meister (UC Santa Barbara) and Ben Ramirez (California Institute of Technology). Reviews of the first meeting were glowing. Monitors John Law and Alan Cowley recommended not only that the Bioinorganic Chemistry GRS continue but also that GRC consider proposals for student-run additions to other Gordon Conferences. Law and Cowley should be happy to know that participation in the Bioinorganic GRS grew from 62 attendees in 1996 to 102 in 2004 and that two new Graduate Research Seminars, Molecular Energy Transfer and Analytical Chemistry, took place in 2005. The success of the Bioinorganic GRS is also demonstrated by its longevity: GRC's seventy-fifth anniversary coincides with GRS's tenth!

(continued)

Harry B. Gray and John S. Magyar
California Institute of Technology

A *Lively* Field from *Lively* Meetings

John Magyar

From the beginning, students attending the Bioinorganic GRS have benefited from the strong support of both the GRC organization and MIB participants. The bioinorganic chemistry community that has developed over the years at MIB warmly welcomes the students. Everyone is energized by the incredible infusion of enthusiasm that arrives with the influx of young scientists. Thursday evening at MIB is a joint MIB-GRS session, with dinner followed by a talk from a distinguished bioinorganic chemist. The evening poster session is a "show and tell" by the student participants. The entire evening has become a terrific opportunity for students to meet and interact with established researchers, and vice versa! This interaction can be extraordinarily valuable for young investigators. Discussions I had as a student at these meetings have led to critical insights, new interpretations, several exceedingly fruitful collaborations, and many new friends.

Most, but not all, of the MIB conferees depart on Friday morning. A few remain, either as discussion leaders or simply because they enjoy the student conference. (Several professors have remarked after their first GRS experience that in the future, they would consider skipping MIB entirely and only attend the student meeting!) From Friday on, all the speakers and most of the participants are grad students or postdocs. The atmosphere is not at all intimidating, leading to lively discussion and broad participation, and students get to know their peers in bioinorganic chemistry, people who will be their colleagues for decades to come.

The strong support of the MIB community for the GRS is also evident in other ways. As 2004 GRS chair, I was thrilled that nearly every discussion leader I invited immediately and enthusiastically agreed. When I asked professors to suggest students to speak, I was inundated with recommendations. In addition, the MIB chair is a mentor to the GRS chair, a relationship that can be very valuable to a young first-time meeting organizer.

The huge MIB participation in the Thursday-night poster session demonstrates both the vibrancy of the bioinorganic community and its strong support for young investigators. Every year both the poster room and the hallways are packed with conferees in highly animated discussions of science. The Gordon Conferences played an important part in the birth of bioinorganic chemistry, and they continue today to sustain the growth of this field.

In 1951, as a twenty-eight-year-old scientist, I attended the first Steroid Chemistry Gordon Research Conference. At the meeting I found myself among top-notch chemists responsible for the rapid advances of steroid chemistry over the past half-decade, in a setting much like the Olympic arena where world-class athletes perform, compete, and demonstrate their brawn.

My own path to the height of the organic chemistry field had been steep and fast. As the offspring of two European physicians, my interest in medicine from a young age was not surprising; however, as a teenage refugee in the United States, I was financially unable to attend medical school. Learning of the many pharmaceutical companies in New Jersey, I sent employment inquiries to every address I could find. The only response came from the American branch of the Swiss company CIBA. Starting work there in 1942, immediately after graduation from Kenyon College, was my first foray into the world of medicinal chemistry.

As a junior chemist I was assigned to the lab of Charles Huttrer, where success came at an unusually fast pace. In less than a year we developed one of the world's first antihistamines—Pyribenzamine, or tripelennamine. My interest in steroids was inspired by older laboratory colleagues at CIBA and by Louis Fieser's treatise *Natural Products Related to Phenanthrene*. Recognizing the difficulty of earning a Ph.D. while working full-time, I left CIBA in 1943 to attend graduate school at the University of Wisconsin at Madison. I arrived just in time to witness Alfred Wilds's and William Johnson's groundbreaking work on the total synthesis of steroids. (Only one steroid, equilenin, had been totally synthesized at the time.) I was inspired by Wilds and set the sights of my doctoral thesis on the transformation of testosterone into estradiol.

COMPETITION, STEROIDS, & MAPLE SYRUP

Racing through my graduate studies, I received my Ph.D. in 1945 and returned to CIBA to continue work on antihistamines and antispasmodics. I used the independent research time CIBA granted to its chemists for steroid research just as the field's breakthroughs were erupting. By 1949 Philip Hench and his colleagues at the Mayo Clinic had discovered the antiarthritic properties of cortisone. Millions of unrelieved arthritis sufferers seemingly guaranteed a huge market, but the partial synthesis of cortisone from bile acids (achieved by Lewis H. Sarett at Merck) had so far proven time consuming, complex, and expensive. Later that year I accepted an offer from George Rosenkranz to join the tiny company Syntex in Mexico City as associate director of chemical research. Our goal was to develop a superior synthesis of cortisone from a plant raw material (diosgenin from Mexican yams).

Though our methods were different, the Syntex team was engaged in close competition with two groups at Harvard University—one run by Robert Burns Woodward and one by Louis Fieser—and a group at Merck under Max Tishler. In June 1951 our three-man team sent a communication to the editor of the *Journal of the American Chemical Society (JACS)* titled "Synthesis of Cortisone." We later learned that the journal (edited at Harvard) had agreed to hold up publication for a few days until Woodward and Fieser had submitted their communications. The Merck group's submission was accepted shortly after. Competition was fierce; nonetheless, four communications appeared in the August issue of *JACS*, printed with clear indication that our submission had been accepted first—then a psychologically significant point that a half-century later seems rather childish.

The 1951 Steroid Chemistry GRC brought face-to-face all the competitors in the race to synthesize cortisone, as well as other chemists and biochemists doing steroid research. One would imagine this to be a rather contentious and awkward event. But as I wrote in *The Pill, Pygmy Chimps, and Degas' Horse,*

> [a]lthough my story thus far would seem to illustrate the competitive aspects of scientific research, it does not disclose the collegiality and mutual esteem that even bitter competitors, like athletes or mountain climbers, display after an event. . . . One afternoon, the leaders of the various cortisone teams were reminiscing about the events of the past year. Our competitiveness and paranoia had been forgotten.

Our macho sense of humor, however, had not. My Wisconsin classmate Gilbert Stork (then at Harvard), Woodward, Sarett, and I concocted a communication to the editor of *JACS* titled "Synthesis of Cortisone from Neohamptogenin"—the latter a supposedly inexhaustible steroid precursor from New Hampshire maple syrup. In the end we got cold feet and did not try to get it by the Cerberus guarding the *JACS* gates.

For the next three years I was graced with the honor of chairing the conference. We abided by the Gordon Conference code that fostered collegiality and pioneering scientific work: each conference had fewer than a hundred scientists participating; each participant had his moment to strut his scientific chops, and then happily obtain the pummel of tough criticism he must absorb to obtain recognition from his colleagues; and time was always taken to breathe in lakeside air and engage in the human act of socialization. (I deliberately use the male gender since, to my shame, I do not recall a single female speaker!) The Gordon Conferences magically bring out the best in scientists whose imaginations are defined by their research and whose prides are entrenched in their achievements.

Carl Djerassi
Stanford University

Intellectual
Fountain of Youth

Frances A. Houle
IBM Almaden Research Center

When I began work in 1980 at the lab that is now IBM's Almaden Research Center, I was asked to consider how to study reaction mechanisms important to the formation and processing of materials for semiconductor chips. The industry was in transition: manufacturing processes were shifting from the use of various liquid solutions to all-dry processes stimulated by radiation and charged particles, and there was little understanding of how any of it worked at a fundamental level. Physicists and electrical engineers were making good progress in developing the processes used for making chips without much formal chemistry training. At the same time the electronics industry began hiring chemists to work in R&D and manufacturing, and we began to delve into the very new chemistry of these processes.

Working in an interdisciplinary environment was exciting and challenging, but we chemists were quite isolated. We grew adept at speaking the language of our physics and engineering colleagues, but we knew few with whom we could talk about the chemistry we found so interesting. Academic departments of chemistry and chemical engineering and the major scientific societies were focused mainly on traditional fields, so there was no forum at which to present our papers and discuss our work. In response to this situation Len Interrante (then at General Electric and now at Rensselaer Polytechnic Institute) and the late Bruce Scott (of IBM's Thomas J. Watson Research Center) had the idea to start a Gordon Conference for chemists working in the electronics industry.

The first meeting of the Chemistry of Electronic Materials GRC was held in Santa Barbara, California, in the winter of 1985. The conference was tremendously broad and interesting, covering topics from photolithography to surface science. I vividly remember sitting in the back of the room, looking around at the audience, and marveling that there were so many of us there. I saw a lot of chemists happy to talk with one another, but I wondered if our shared training alone would suffice to sustain the conference. In subsequent meetings the conference chairs and steering committees worked hard to define who the conference should serve and to build a community of attendees who identified themselves as electronic materials chemistry researchers. Over time the conference has become a multidisciplinary, vibrant meeting, bringing together people from universities, national laboratories, and industrial research centers worldwide and addressing a variety of emerging themes.

I have no doubt that the GRC organization has been central to the establishment of the field of electronic materials chemistry. The conference's attendees and speakers over the years constitute a "who's who" of the field's major contributors. Organizers of meetings sponsored by scientific societies now regularly hold sessions on aspects of electronic materials chemistry, and many academic departments have materials chemists among their faculty. I chaired the 1994 conference and attended the conferences until 1999, when my own research took me deeply into photolithography. Interestingly, photolithography is not a major focus of the contemporary Chemistry of Electronic Materials Conference despite having been on the program of the first meeting. Since then I have been attending nanofabrication conferences, including the Nanostructure Fabrication GRC.

The Chemistry of Electronic Materials is an example of a Gordon Conference founded by industrial chemists who needed a venue to discuss the leading-edge topics in their field. But Gordon Conferences play a much broader role in the overall vitality of industrial research. Unlike that of university labs, the staffing of industrial labs tends to be rather stable, with only a few new people coming in each year. This stability promotes efficient teamwork among experienced professionals with backgrounds in many disciplines, but it does little to bring new perspectives to the laboratory. The agenda of the typical conference attended by industrial chemists is usually rushed and packed, with very short talks. This format is great for the transfer of information but provides neither the time or space for the reflection and discussion crucial in innovative science nor the opportunity to meet new people and start new collaborations. Gordon Conferences, however, provide a forum to slow down and focus on new research, learn about areas of science not represented by our own institutions, and think broadly and imaginatively. In other words, GRC is a fountain of intellectual youth.

A Fine Balance

Paul C. Lauterbur
University of Illinois at Urbana-Champaign

The value of the Gordon Research Conference series is its fine balance among various scientific fields, between pure research and application, between experts and newcomers, and between scientific discussion and free time for afternoons on the lawn or tennis courts and boisterous, convivial evenings in the bar.

I have attended more Gordon Conferences than I can easily remember: everything from Magnetic Resonance, to Magnetic Resonance in Medicine and Biology, to Isotope Effects, to Inorganic Chemistry. The variety of conferences I have attended has helped me learn about my own field, meet its leaders, and appreciate new, peripheral fields with a glimpse into the inner thoughts (and backhands) of their practitioners. This exposure has helped stimulate new thoughts and bolster my confidence in my ability to understand the "cutting edge" in new fields.

Long, lazy summer afternoons framed the intense morning sessions, providing experiences that all scientific meetings should procure. I shall always treasure my memories of Gordon Conferences and the glow they cast over the rest of the year.

Midwife to the Birth of
Organic Electronics

Charles B. Duke
Xerox Innovation Group

In 1972 I left a fine job as a professor of physics at the University of Illinois at Urbana-Champaign to come to the Xerox Corporation in Webster, New York. Xerox was in its heyday: at the time it was the fastest-growing firm in the history of the New York Stock Exchange. But troubles were on the horizon. A consent decree in 1975 opened the company's patent portfolio to all comers, and in the three years since I had arrived, its stock had fallen from $172 to $50 per share.

My hiring manager was assigned the task of starting a new chemistry laboratory in Canada (which became the Xerox Research Centre of Canada in 1974), and I was assigned his job of managing materials research in Webster, as well as planning the technology futures for copier-duplicator businesses. The creation of a new generation of xerography on which Xerox could obtain new patent protection became a prime focus of both efforts, because it would allow Xerox to stave off the competitive assault from Eastman Kodak and IBM in the United States and from numerous Japanese firms attracted by the consent decree. The effort to create a new generation of xerography led to the birth of organic electronics as a new technology-led industry. The path to this new industry ran straight through the Gordon Research Conferences.

The xerographic process involves several steps: charging an insulating drum; creating a charged image by exposing the drum to light; developing the image with charged, pigmented powder (toner); transferring this powder image to paper; and then fusing the image onto the paper with heat and pressure. First- and second-generation xerography, the subjects of Xerox's patents before 1975, used metal drums or belts coated with amorphous selenium or arsenic-selenium alloys.

For the third generation of xerography we envisaged using flexible organic belts. They would be cheaper, and we would be able to wrap them around small rollers to form flat areas suitable for full-frame flash exposure. This would permit high-speed copying using small console machines instead of large machines that embodied massive drums or belts. But there was one small problem: the flexible organic materials of the day trapped photo-injected charges, and the resulting photoreceptors had very limited lives.

The solution was a drive to develop organic materials for electronic applications. The materials would have controllable electrical properties and would behave like familiar network inorganic semiconductors but would be flexible, cheap, and easy to fabricate. At the same time major universities were engaging in research on "quasi one-dimensional" crystalline semiconductors, most of which were organic. Xerox also started a research program in this area to couple our organic solid-state research with this vibrant academic effort.

The two streams of activity came together in 1977 with back-to-back conferences. The Chemistry and Physics of Solids Gordon Research Conference met in Plymouth, New Hampshire; immediately following, a special New York Academy of Science conference called Synthesis and Properties of Low-Dimensional Materials met in New York City. Dwaine Cowan, of Johns Hopkins University, and I co-organized the GRC conference, and two other Xerox scientists, Joel Miller and Art Epstein, organized the one in New York. Previously, research efforts into the synthesis, theory, and properties of electronic organics had been separated, but these meetings brought them together as a coherent group. They also brought together participants from a variety of firms, academic institutions, and disciplines that had seldom participated in the same forums in the past.

Within a few years this rich mix of people and ideas led to two important breakthroughs. First, it became evident that the organic polymers that best exemplified semiconductive properties were not organic semiconductors in the context of solid-state energy band theory. Rather, they were best described using disorder-induced localized state models, which predicted design rules for photoconductors and toners that were dramatically different from the rules based on energy band theory in use at the time. Within a decade the new models led to the design and fabrication of large-area photoreceptor belts and an organic xerographic materials industry that generated a billion dollars per year. Second, the notion of doping organic polymers with atoms and small molecules (adding materials to facilitate conductivity), a practice already prevalent in the photoconductor industry, was extended to other (degenerate ground state) polymers. This in turn created the field of organic synthetic metals, for which Alan Heeger, Alan MacDiarmid, and Hideki Shirakawa won the Nobel Prize in chemistry in 2000.

The Organic Thin Films and Surfaces GRC I organized in Ventura, California, in the winter of 1983 further extended the momentum created in 1977. Numerous applications of organic electronics were explored at this conference—including large area displays, plastic batteries, semiconductor processing, solar energy, optical storage, electro-optics, photography, and electrophotography—by a diverse group of industrial and university scientists who seldom met in other forums. Chemists and physicists mingled intimately. Alan Heeger led a session on organics in energy and electrochemistry. The discussion topics presaged contemporary efforts in organic light-emitting diodes, flexible displays, and printed organic electronics.

Unfortunately, there was not enough economic incentive for industry to invest in the production technologies for organic electronic applications other than photography and electrophotography. Few people realize that xerographic photoreceptors are as complex as large-scale silicon integrated circuits but are fabricated by the square mile rather than by the square centimeter. But the researchers did not give up; in fact, many of the players in the early Gordon Conferences in this area are still prominent leaders in the development of organic electronics materials and devices.

The birth of organic electronics illustrates the immense intellectual and economic benefits that result from a free flow of technical information. Individual scientists gained access to model concepts, experimental measurement techniques, and methods of materials fabrication practiced in only the most advanced academic and industrial laboratories of the day. They encountered opportunities for new collaborations that often took the field in unanticipated directions. Xerox and other firms obtained insight into and validation of design principles that yielded specific copier products and materials of considerable economic value. This led to industry growth: new jobs and capital investment spread widely throughout the associated business communities. The Gordon Conferences facilitated these results by providing forums for off-the-record exchanges organized and led by both industry and academic scientists. The openness and intensity of Gordon Conferences aided enormously in creating intellectual and commercial value. GRC was the midwife at the birth of the organic electronics industry, contributing directly to both the intellectual and economic vigor of the United States.

Desperately Seeking
ELECTRIC FANS

Anthony Czarnik
University of Nevada at Reno

Somehow I managed to miss all notice of the Gordon Research Conferences as a graduate student and postdoctoral fellow. In retrospect, it is hard to imagine how that could have happened. I suppose I missed the meeting announcements because I had given up my subscription to *Science*. In any case, in 1983, when I became an assistant professor of chemistry at Ohio State University, my friends David Hart and Matt Platz told me that the meetings were great. I think David helped get me an invitation to speak at the 1984 Medicinal Chemistry GRC at Colby-Sawyer College in New Hampshire.

I was enormously excited to have the opportunity to give my first major presentation to a group of such well-known chemists. The whole experience—perhaps excluding the heat and the fact that I lacked a fan in my room—was wonderful. I soon became a regular attendee at the Bioorganic Chemistry GRC. Every meeting was valuable: the good talks were stimulating, and even the talks I found less interesting provided time to let my mind wander.

At one Bioorganic Chemistry meeting in the mid-1990s I found myself in a discussion with several other attendees about one of my pet peeves—why cell biologists who needed fluorescent indicators for intracellular analytes did not seem to be aware of the work being done by chemists who were synthesizing them. This conversation led directly to my venting in the journal *Chemistry and Biology*. I wrote a piece titled "Desperately Seeking Sensors" in twenty-four hours; as it turns out, it is still one of my most cited papers.

I attended the Bioorganic Chemistry conferences long enough to be elected chair for the 1997 meeting, a duty I shared with John Griffin. I was so proud to have had the opportunity to co-organize a meeting; I really felt I was giving back to a conference that had given me so much.

In early 1998 GRC management put out a call for proposals for new meetings that might draw attendance by more industrial scientists. Early Gibson Island Research Conferences organized by Neil Gordon brought together many industrial chemists who could, by signing nondisclosure agreements, talk about the problems in their current research projects and get help solving them. Over the years academics had come to make up a larger and larger percentage of attendees, presenters, and organizers. Thus, initiating a few new meetings that spoke to industrial science as in Gordon's time seemed like an excellent idea.

I had just been selected as the editor for the American Chemical Society's new *Journal of Combinatorial Chemistry*, and I was confident that combinatorial chemistry was the kind of topic GRC management had in mind. Hervé Bouchard felt the same way and had submitted a similar proposal; so Hervé and I served as cochairs of the first Combinatorial Chemistry (CC) GRC in 1999. I worked extremely hard to raise money for the meeting (starting a meeting from scratch felt like a huge responsibility). My instinct told me that companies would perceive the relevance of the Combinatorial Chemistry Conference to their own operations in medicinal chemistry, and I was proved correct. We were supported generously (which was fortu-

A Stereochemical Merger

Ernest L. Eliel
University of North
Carolina at Chapel Hill

ate, because I was able to purchase three large fans for the poster room to give relief from the beastly heat that summer at the Tilton School in Tilton, New Hampshire; I later donated the fans to the school as a "thank you" for hosting us that week.) The meeting was oversubscribed and was such a success that Carlyle Storm suggested we start meeting annually, on the condition that we agree to meet in 2000 at the site in Il Ciocco, Italy. With little to no persuasion required, the attendees voted a resounding "yes!" By alternating between U.S. and European sites, the Combinatorial Chemistry Conference has managed to retain high levels of interest from chemists on both sides of the Atlantic.

Perhaps because things had gone so well for the first Combinatorial Chemistry Conference, I was invited to serve as a member of the GRC Selection and Scheduling Committee starting in 2000. I have found that being a part of the group that makes life-and-death decisions for meetings is a heavy, but essential, charge: meetings must change as emphases in science change. Attending annual meetings with Carlyle, Nancy Ryan Gray, and the twenty or so scientists committed to preserving the unique atmosphere of GRC meetings has been a privilege. This group takes every aspect of the Gordon Conferences seriously, and discussions, even when heated, are always respectful. GRC is a great organization supported by a dedicated staff, and it deserves to be sustained by all who have reaped its many benefits.

Stereochemistry has its origins in the middle of the nineteenth century with chemists Louis Pasteur, Joseph Le Bel, and Jacobus Henricus van't Hoff. The field was very active in the second half of the nineteenth century but declined in activity in the first half of the twentieth. Stereochemistry changed again around 1950 with the pioneering work of my mentors: Derek H. R. Barton, who worked on conformation, and Vladimir Prelog, whose research centered on configuration. In 1962 I published the first modern English text on organic stereochemistry, followed by four related coauthored books and a series called *Topics in Stereochemistry* coedited by Norman L. Allinger and first published in 1967.

In 1965 I attended the first of what are now the annual EUCHEM Conferences on Stereochemistry (or the Bürgenstock Conferences) held in an elegant hotel on a bluff overlooking Lake Lucerne in Switzerland. Only a few Americans are invited to these yearly conferences, and so I felt it desirable to initiate a Stereochemistry Conference in the United States. This plan had the full support of my fellow stereochemist Kurt Mislow (Princeton University). However, a complication arose: in 1975 James Morrison (University of New Hampshire), Donald Valentine (Hoffmann–La Roche), and Albert I. Meyers (Colorado State University) initiated the Asymmetric Synthesis Gordon Conference. Morrison chaired the 1975 conference and Valentine the next one in 1978. Both Mislow and I felt that this was too narrow an area for a sustained conference, but we also realized it would not be acceptable to have a second Gordon Conference in the general area of stereochemistry. Thus, with the active help of Al Meyers and Gordon Conference board member David Lemal, we persuaded the originators of the Asymmetric Synthesis Conference to merge it into a broader Stereochemistry Conference, with the 1978 conference first taking the name and myself taking on the responsibilities of vice chair.

(continued)

A Stereochemical Merger

Of the twelve talks at the 1975 conference eleven were on stereoselective synthesis. By 1978, with some input from the vice chair, only seven of the thirteen lectures were on that subject. By 1980, when I was chair, the incidence of stereoselective talks had dropped to four out of thirteen, as it was my intent to present a very broad view of the field. In 1982, under the chairmanship of Al Meyers, the percentage rose again to almost half and has stayed at near half ever since. This is probably as it should be, for although stereoselective synthesis is not the be-all and end-all of stereochemistry, it is probably the subject of greatest interest to the substantial number of attendees from the pharmaceutical industry (which has supported the conference financially over the years).

Poster sessions were first added to the 1980 conference; eighteen posters were presented, and posters have been included ever since. Generous financial support from the Gordon Conference board over the years made it possible to defray conference fees for a number of young attendees, in addition to the lecturers and discussion leaders. Total attendance in 1980 was 123; in the last two decades it has fluctuated between 80 and 129.

The Stereochemistry Conference now takes place every two years. Occupation of the chair has generally oscillated between industrial and academic chemists. In addition to stereoselective synthesis subjects now cover such broad areas as nuclear magnetic resonance spectroscopy, optical rotation and circular dichroism, molecular mechanics, inorganic and bioorganic stereochemistry, crystallography, macromolecular chemistry, and other subjects, including green chemistry. In 1986 and 1990 molecular modeling workshops were also held. The breadth of the conference is its strength, since it brings together chemists from many areas to inject new points of view.

From 1975 to 1986 the Stereochemistry Conference was held at Plymouth State College in New Hampshire's White Mountains, and afternoons were often devoted to mountain climbing or swimming. In 1988 the conference moved to more comfortable quarters at Salve Regina College in Newport, Rhode Island, where afternoons tend to be divided between recreation and poster sessions or workshops.

I have attended all but one of the Stereochemistry Conferences in the last twenty-five years and have always found them interesting and enjoyable.

Gordon Research Conferences continue to be among th most stimulating and rewarding of scientific gathering GRC provides dedicated meetings for well-establishe active fields and was one of the first organizations to cr ate conferences (to which new participants are attracte for emerging, often multidisciplinary research areas. the 1960s two subjects were centers of increasing activ ity: photochemistry and magnetic resonance. Stimulate by the Photochemistry Gordon Conference that began 1964 (known as Organic Photochemistry until 2001) an the Magnetic Resonance Gordon Conference that starte in 1958, both of these fields were extended by multidisc plinary efforts. Individually, both of these meetings wer of great value to a number of scientists at Bell Telephon Laboratories, where I worked at the time. Together, th two series helped develop a unique research focus on th combination of photochemistry and magnetic resonanc

In photochemistry connections were being mad between the excited states of starting molecules ar reaction products. The earlier, largely empirical approac of the field (as developed by organic chemists) wa changing into a combination of organic and physica chemistry, especially in the use of spectroscopy. The lak oratories of George Hammond at the California Institute of Technology and of George Porter at the University of Sheffield and the Royal Institution in England playe major roles in the early Photochemistry Conferences, th first of which was chaired by Tony Trozzolo of Bell Lak oratories. The opportunity to learn from and teach on another in the mutually supportive atmosphere of thes conferences facilitated rapid progress. Newcomers to th science could consider how recent developments in phc tochemistry could be applied in their own fields.

Scientific *Synergism*

Edel Wasserman
DuPont, retired

The study of magnetic resonance in condensed phases effectively began in the late 1940s as a discipline within physics. During the 1950s chemically distinct nuclei and their neighboring atoms could be identified; this information was of great importance for chemists. At the Magnetic Resonance Conferences chemists learned from physicists the fundamentals of the field and the wide-ranging techniques that were becoming available. Anatole Abragam, Nicolaas Bloembergen, Ray Orbach, and John S. Waugh were among the leaders who participated. The chemists could apply their newly acquired knowledge to larger molecules, while the physicists acquired new structures and phenomena to study. The cross-disciplinary fertilization stimulated by these conferences was beneficial to both groups and has continued for five decades.

In the 1960s Bell Labs was a major center for nuclear magnetic resonance (NMR) and the largest customer of Varian, Inc., the primary manufacturer of NMR instrumentation. Dave McCall, Bill Slichter, and Dean Douglass studied organic polymers at Bell Labs; Bill Yager, a physicist who had pioneered in the paramagnetic resonance examination of organic free radicals, was a few doors away. Located along the same corridor at Bell, Bob Murray, Gerry Smolinsky, and Tony Trozzolo were studying carbene and nitrene chemistry. Missing two bonds with hydrogen, many of the carbene and nitrene species had two unpaired electrons and were ground-state triplets. Independently, Trozzolo and I had developed interest in their electronic structures and how to detect them. Valued theoretical input came from Larry Snyder and Si Glarum. Following an unexpected but thoughtfully pursued paramagnetic resonance observation by Yager, we developed a facile and efficient route for obtaining the most important triplet parameters. Months after we completed our first studies, we learned that physicists had been using similar methodology primarily in the study of inorganic structures. Even with contributions from the Gordon Conferences, communication across fields was imperfect.

This combination of photochemical generation of triplet systems with paramagnetic detection was unusually productive and continued over the next decade in several different directions. One highlight was elucidating the now generally accepted structure of the parent carbene (CH_2). Another was the formation and study of di- and tri-carbenes and nitrenes, now often referred to as "organic ferromagnets" because of their high spin states.

For those of us who attended the conferences, there was an additional bonus: on returning to Bell Labs we continued discussions with colleagues. Chatting with them and with other preeminent practitioners was a major activity, and we became each other's teachers and students. An electric atmosphere existed at Bell and at some other industrial laboratories. The stimulation yielded combinations of existing fields, which became new disciplines. Unfortunately that era of basic industrial research has passed. But new trends in academia have led to interdisciplinary centers and industrial-academic collaborations with highly productive environments. These will be critical for the future of science and for industry.

time to
COME CLEAN

David A. Tirrell
California Institute of Technology

I begin with a confession: I attended my first Gordon Conference lecture "illegally." I was a first-year graduate student at the University of Massachusetts in Amherst, and I noticed a lecture on the program of the Polymers Gordon Conference that I really wanted to hear. I did not know anything about Gordon Conferences, but I did know that I was not registered for the meeting. I drove up to New London, New Hampshire; found Colby-Sawyer College; and figured out where the lecture room was. My plan was simple: the lecture I was interested in was scheduled right after the morning break; I would wait until everyone returned to the lecture room and the lights went down, and then I would slip in the back door. Surely no one of consequence would be sitting in the back of the room, and I could slip out again, unnoticed, as the lecture ended.

I was almost right. The lights went down, I slipped in, and I quickly found a stool (yes, a stool) against the back wall. Because my eyes were not yet adjusted to the dark, I could not see anyone around me for a minute or so. As my neighbors gradually materialized, I noticed that everyone (except me, of course) was wearing a name tag—a very large and very visible name tag that identified them as legitimate conferees. As I looked more closely at my neighbor to the left, I made out his name: Walter Stockmayer! Convinced that my career in polymer chemistry was over before it had begun, I turned slightly to my right and listened as best I could to the lecture I had come to hear. In the end I did slip out, probably unnoticed, and returned to Amherst. I remember almost nothing about the lecture, but I remember vividly my first encounter with Walter Stockmayer more than thirty years ago.

My second encounter with Professor Stockmayer also came at the Polymers Conference a few years later. This time I was an assistant professor at Carnegie Mellon University (CMU) and properly registered for the meeting. This was one of those interactions that the Gordon Conferences are famous for and that probably would never have occurred anywhere else. We were leaving the lunchroom, about to enjoy a free summer afternoon. "Stocky" asked the usual questions about what I wanted to do in my research program at CMU, and his interest seemed so intense and so genuine that I left with a new sense of confidence that perhaps there was something worthwhile in my research.

Another such meeting happened at about the same time. Norbert Bikales was the director of the polymers program at the National Science Foundation and a regular contributor to the Polymers GRC. Norbert made a point of seeking out the young people in the field to be sure that he knew what they were doing and that he could provide whatever support was appropriate. Norbert was helpful to me in many ways over the years, but what I remember most clearly about our conversation at GRC was his asking, "How long do you think you will work on this project?" I had never considered such a question; I was just thinking that the work would go on indefinitely. In fact, it went on for only a few more years, and my laboratory was soon moving in different directions.

The Stories *of* Generations

One aspect of the Gordon Conference operation that has served the macromolecular chemistry community (and the broader scientific community) well is the willingness of GRC leadership to start new conferences when they are justified and to terminate them when they are no longer needed. For example, polymer chemists and physicists began moving into many new areas in the 1970s as interest grew in various kinds of special polymers, and new Gordon Conferences provided critically important venues for discussion of these new research directions. But conferences must also show evidence of real vitality in order to justify continued GRC support. The willingness of GRC leadership to demand evidence of excellence, to seek advice and criticism from the scientific community, and to subject conference activities to continuous, candid review has been a key element of the remarkable seventy-five-year history of the Gordon Research Conferences.

I grew up amidst the infectious enthusiasm scientists exude for Gordon Research Conferences. Often my dad would bring home foreign visitors after the Photosynthesis Gordon Conference, while I sat in the background, rather unnoticed, rapt by the intense and lively discussions.

When I started as an assistant professor at the University of California at Santa Barbara in the mid-1980s, the proximity to the winter Gordon Research Conferences—initially held in Montecito at the Miramar Hotel and later in Ventura and Oxnard—was really convenient. I was working on a new metalloenzyme, vanadium bromoperoxidase from marine algae, and was particularly interested in bioinorganic chemistry. I had already been to the Metals in Biology (MIB) Gordon Conference as a postdoc in 1984 and was familiar with the incredible caliber of the talks and the exceptionally vibrant and animated atmosphere. The breadth and depth of the expertise at this GRC, above all others, have shaped my scientific thinking immensely. In fact, I have applied to every MIB Conference since that first time.

Alison Butler
University of California
at Santa Barbara

But early on, my research took a turn toward biosynthesis of halogenated marine natural products, so the Marine Natural Products Chemistry (MNP) Conference was the obvious place to turn. From the beginning I was captivated by the attendees' descriptions of their travel adventures to far-off reaches of the world's oceans to collect samples, as well as by the new compounds and enzymes that they reported. Through collaborations established with Bill Fenical and John Faulkner and their students at MNP, I went on several research cruises that screened marine organisms for halogenating enzymes. But what really caught my eye in the late 1980s was the emerging work presented at the MNP Conference on culturing marine microorganisms and the kinds of bioactive compounds that were isolated. My research group delved into isolating compounds from marine bacteria, although our focus was (and still is) on compounds that coordinate transition metal ions, such as siderophores. After giving a talk at the MNP Conference in 1998 on the unusual photoactive and amphiphilic marine siderophores we were discovering, and trying to convey the importance of inorganic chemistry and bioinorganic chemistry to the group of largely organic chemists, I was at first astounded to be asked to run for the chair position of the GRC and then very humbled when I was actually elected. In the end organizing the MNP Conference was a tremendous experience; I read widely in the marine natural-products literature and talked to everyone I could think of about what was new and exciting before putting together the program.

(continued)

Two years later when I was speaking at the MIB Conference about the photoreactive and self-assembling amphiphilic siderophores we had discovered, I thought I would try to convey the importance of the extreme and unique transition-metal composition of the world's oceans on the likely discovery of new bioinorganic chemistry in this environment. (After my talk one person in the front row actually held up a sketch of a baseball flying through the air with the words "Home Run" scrawled beneath, which was a huge relief.) The ensuing discussion—which ranged from molecular aspects of the photoreactive iron(III) siderophores to global implications of the photoreactive cycles in the ocean—remains crystal clear in my mind and continues to shape the evolution of my ideas. After this talk I was asked to run for MIB Conference chair and was thrilled (if not also humbled again) to be elected. Organizing the 2004 MIB Conference was another extremely satisfying experience in terms of delving into new areas of metals in biology, talking to many friends in the MIB community, and then sitting back during the conference and watching not only the talks unfold but especially the discussions.

Now I am in the throes of organizing yet a third conference, the Environmental Bioinorganic Chemistry (EBIC) Gordon Research Conference for the summer of 2006, with Bradley Tebo as cochair. You might ask, is she a GRC chair junkie? No! I even turned down the initial invitation to run for the chair position, but by the time of the business meeting of the first EBIC Conference, which was held in 2002, I was so inspired that I knew I wanted to help be a part of its success. This new GRC was conceived by Ed Stiefel and François Morel as a spin-off from the success of our National Science Foundation–funded Center for Environmental Bioinorganic Chemistry (CEBIC) that is headquartered at Princeton University. But it is also a direct reflection of the sustained success of the MIB Conference and the interdisciplinary nature of this field.

What is it about Gordon Research Conferences th[...] inspires the affection so many of us feel for this confe[...] ence series? Primarily, the science is outstanding. We ta[...] the motto "in the spirit of the Gordon Conference" to he[...] by exposing our newest results and latest hypothes[...] (after all no one can quote us) and synthesizing new ide[...] from the discussions. We also work hard and play hard, [...] the point of exhaustion, but our students and colleagu[...] clearly "get it" when we arrive back home eager and r[...] ing to go. Importantly, Gordon Conferences give young[...] scientists one of the best environments to meet and inte[...] act with top scientists in a field, in a way that rarely ha[...] pens at larger conferences. In this regard the Gradua[...] Research Seminar GRC that overlaps with the Thursda[...] evening session of the MIB Conference is a wonderf[...] avenue for graduate students and postdoctoral fellows [...] interact with the MIB crew.

Given the scope of science covered by the GR[...] series, our scientific community can be surprisingly sma[...] and close-knit. I never really considered my dad's resear[...] to be close to my own; yet I have often met people wh[...] had attended Gordon Conferences with him, and the st[...] ries picked up from where they had left off years ago. Ev[...] in my own home now GRC stories captivate the interes[...] of my young daughters, who have helped set up th[...] Sunday-night chair's receptions for two Gordon Confe[...] ences. Once they even added their own poster of some [...] their lab work to a poster of mine. Now many decad[...] after I first heard of Gordon Conferences, I see that th[...] infectious enthusiasm that scientists at Gordon Confe[...] ences show for their research and for each other is ru[...] bing off on yet another generation.

Illustrations of
Life

Roy Vagelos
Merck, retired

In the course of my career I was fortunate to have worked in government, academia, and industry, where I eventually became the CEO of Merck. I started on this path in 1956, when I took a postdoctoral position in biochemistry with Earl Stadtman at the National Institutes of Health (NIH). It was an unusual arrangement for both of us: for the first time, Earl was bringing an M.D. instead of a Ph.D. into his laboratory, and in addition to caring for patients with heart disease, I agreed to spend half my time doing research. I dove into biochemistry, starting with a textbook that Earl gave me, and in 1958 Earl set me on a path of independent research into fatty acid metabolism.

In 1963 I was invited to my first Lipid Metabolism Gordon Research Conference. It was an incredible honor to share the podium with senior researchers, including Konrad Bloch of Harvard University, Irving Fritz of the University of Michigan, and Salih Wakil of Duke University. There I was, a pipsqueak only a few years out of a postdoctoral fellowship, speaking alongside these giants of biochemistry. In 1965 I presented alongside Wakil and Feodor Lynen of the Max Planck Institute in Germany, who was a buddy of Earl's and shared the 1964 Nobel Prize with Konrad for work on cholesterol synthesis. Bloch, Wakil, Lynen, and I were all competing to understand fatty acid biosynthesis. In 1953 Fritz Lipmann, with whom Earl had worked as a postdoctoral fellow at Massachusetts General Hospital, won half the Nobel Prize in physiology or medicine for discovering coenzyme A (CoA), a molecule critical to fatty acid metabolism. Lynen showed exactly how CoA was involved in fatty acid breakdown. Wakil and I simultaneously but independently discovered the acyl carrier protein, which my team later found was the single

protein to which all the intermediates of fatty acid synthesis are bound. We were ardent competitors.

This collection of individuals demonstrated the greatest strength of the Gordon Conferences: they operated without hierarchy. Attendees, including young biochemists like myself, were selected by chairs on the basis of the exciting research they were doing that year. It was amazing to sit with people who were significantly older and much more accomplished and be in a position to swap information with them. I come from a modest background and have always had to work hard for my achievements, so it was both inspiring and intimidating to be competing eyeball to eyeball with senior researchers. The Gordon Conferences allowed you to get to know the work of leading scientists and also to get to know them personally while you were eating together, fishing, or socializing. For example, I played a lot of tennis with Harvard biochemist Eugene Kennedy. Even when you were in the heat of competition with someone, like I was with Konrad, there was still always a sense of collegiality and camaraderie. Konrad and I got to know each other well, and he later invited me to be a visiting professor at Harvard. Perhaps most important, however, Gordon Conference meetings made me realize that our NIH research group was capable of moving our work along as fast as our competitors. We were able to show that acyl carrier protein contains half of the CoA molecule. Despite being young, we had the insight to take big leaps in our work.

In 1966 I moved on to chair the biological chemistry department at Washington University Medical School, and by 1969 it was my turn to chair the Lipid Metabolism Conference. Trading information at GRCs continued to be

ANTS

extremely productive for my work. Because of the relationships of fatty acids to complex lipids and phospholipids, I learned a lot from other researchers in these disciplines. Later I also learned about steroids. Without having planned it, my career moved into these areas. I went from fatty acid metabolism into complex lipids and ultimately into cholesterol research. This got me thinking about the biosynthesis of cholesterol. The idea of using mutant *Escherichia coli* bacteria to modify the genes responsible for unsaturated fatty acid synthesis at the NIH laboratory came from microbial geneticists I met at Gordon Conferences. That led our research group to use sterol biosynthetic enzyme mutations in Chinese hamster ovary cells to prove that cells needed cholesterol. At Merck in the 1970s and 1980s our researchers showed that inhibition of the enzyme HMG-CoA reductase caused cholesterol levels to decrease and thus demonstrated that reduction in blood cholesterol was beneficial to human health. These insights contributed to Merck's development of the cholesterol-reducing drugs Mevacor and Zocor.

There was little premeditation in the course of my career: it just happened that I was at Gordon Conferences where these topics were being discussed. The breadth of my research grew enormously through both the informal and formal GRC sessions. I did not get stuck in fatty acid metabolism all my life in part because I met people from different fields at these meetings. I used this background as the first step in drug discovery at Merck, which opened doors to the path I traveled to corporate leadership.

Making Connections

In 2007 we will celebrate the twenty-fifth anniversary of the Fibronectin, Integrins, and Related Molecules GRC. In 1982 I attended the very first meeting, then named simply Fibronectin, which had been spun off from the Collagen GRC. As a young postdoc I switched from bacterial protein chemistry to mammalian extracellular matrix biology and was still learning about my newly chosen field of research.

The large and impersonal meetings like Keystone and Federation of American Societies for Experimental Biology that I had previously attended were nothing like my first Gordon Conference. The GRC atmosphere was completely different. I loved being able to sit at the dinner table with the leaders in extracellular matrix biology and eavesdrop on their discussions. It is no surprise that I have very fond memories of that conference. I have been a devoted attendee at subsequent Fibronectin Gordon Conferences, and I served as its chair in 1995.

Jean Schwarzbauer
Princeton University

The first Fibronectin GRC covered a wide range of topics—everything from clinical studies to hard-core biochemistry. Subsequent meetings have been focused on timely central issues, and some meeting themes have had a cyclical life. A memorable controversy at the 1982 meeting, for example, was whether fibronectin, which is a very large protein, had an extended or compact structure. Groups from the United States, Switzerland, and the Soviet Union showed data obtained using a newly developed single-molecule electron microscopic (EM) technique. But each group had used a different solution for depositing fibronectin on the EM grids, and the differences resulted in dramatically different pictures of fibronectin. At the time we thought only one of these structures could be correct. As it turns out, each of these original configurations can occur under particular physiological conditions. Today we still do not know quite what fibronectin looks like, and the topic of its structure remains an important and interesting part of the Fibronectin GRC.

(continued)

Making Connections

The conference's central focus has changed from fibronectin to other molecules and processes over the years. The most significant topic change occurred in 1987 at the conference's third meeting, when many of the presenters' talks involved integrins—newly discovered cell receptors for fibronectin and related molecules. Few of us could foresee the impact that integrins would have on biology and medicine. At the time we were more interested in confirming commonality of the bands on protein gels from different labs or determining whether the proteins found in flies were the same as those in chicken or human samples. It became clear, however, that these receptors were setting a new direction for the field. The 1987 meeting changed the face of extracellular matrix biology.

Recurring meetings are advantageous in that they reflect current interests and can predict future issues in a field. The Fibronectin GRC has done just that by being at the forefront of the field. The excitement over integrins in 1987 was followed by other major turning points for the field, including understandings of the link between fibronectin and signal transduction cascades, of crystal structures for parts of fibronectin and integrins, and of knock-out and knock-in phenotypes in mice.

The personal connections and new ideas that I have developed at this GRC have greatly affected my scientific career. At this conference I gave my first major talk as an assistant professor and made my first European connection, which resulted in an invitation to speak in France. Also important, as a resident in the "girls' dorms" of the New England sites, I have made lasting friendships with other women scientists. Finally, the impact of the Fibronectin GRC is illustrated by the large number of scientists who have attended since its first meeting. Many of the faces that appeared in the 1982 conference photograph can also be found in the 2005 photo, albeit with a few more wrinkles.

Like many of my peers, I was first drawn to organic chemistry by the artistic beauty inherent in the synthesis of complex molecules. Once some of the fundamentals have been mastered, the organic chemistry of medicinal compounds takes on an important creative element, which is furthered by the cross-fertilization of ideas that take place at Gordon Conferences.

My Ph.D. adviser at the University of New Hampshire, Robert Lyle, went to the Chemistry of Heterocyclic Compounds GRC every summer. He would regale his graduate students with stories of sitting on the shores of the lakes in New Hampton and drawing molecular structures in the sand. With that inspiration, and following the advice of John Topliss, I applied to and attended my first Medicinal Chemistry GRC in 1975. I was a young scientist at Schering-Plough, following my postdoc years at the University of Michigan and the Squibb Institute for Medical Research. My research had shifted to the investigation of new antiulcer compounds from earlier graduate work on the partial synthesis of the anticancer agent camptothecin and postdoctoral research on molecules that inhibit the biosynthesis of cholesterol.

The Medicinal Chemistry Gordon Conference was well-established by the mid-1970s, and many of the same people attended every year. In fact, the front row of seats at Colby-Sawyer College was unofficially reserved for the conference "elders," who included Murray Weiner, Nat Sperber, and Jerry Weisbach. I was immediately struck by this assemblage of remarkable people—all leaders in pharmaceutical research from across industry and academia. But a great set of younger scientists was also in attendance: Ray Fuller, who invented serotonin uptake

Artistic Beauty, Synthesis, and
Front-Row Seats

inhibitors; John LaMattina, who went on to become president of R&D at Pfizer; and Peter Gund, who had an illustrious career at Merck in the field of quantitative structure-activity relationships and molecular modeling.

Once I attended, I was hooked, and I participated in the Medicinal Chemistry GRC regularly for several decades. I even made it to the 1981 conference, when I was working for Parke-Davis. I took my two sons to New Hampshire while my wife stayed in Michigan, about to give birth to our daughter. Fortunately, I made it back just in time!

The Medicinal Chemistry meeting opened my eyes to the breadth of research going on at other companies and in academic labs. One of the major strengths of this conference is that it does not focus on a single therapeutic area but instead provides the latest and greatest insights into a wide range of areas and promotes the great discoveries that happen at the interface between therapeutic areas. For example, during the mid-1980s we began a project at Parke-Davis to explore adenosine receptors. In effect, we began with the mechanism and then looked for potent ligands (either antagonists or agonists) of the receptors that we could evaluate to identify a range of therapeutic areas. As it turned out, we found incredibly interesting activities in the areas of pain, hypertension, psychiatric disease, inflammation, and lung disease.

More recently, at Pfizer we have investigated a series of compounds that have the same mechanism of action as Neurontin (gabapentin), a drug that was initially used for epilepsy. Interestingly, these compounds are active in neuropathic pain, anxiety, and a half-dozen other disease areas. As these two examples illustrate, successful pharmaceutical research often happens when scientists apply discovery techniques developed in one area

to totally different therapeutic areas. GRC has likewise advanced the field by bringing together scientists from a wide range of research.

Like other Gordon Conferences, Medicinal Chemistry also serves as a leading indicator of research in the field. *Annual Reports in Medicinal Chemistry*, for which I had the privilege of serving as editor-in-chief for nine years starting in 1990, draws its editors from the Medicinal Chemistry Gordon Conference and begins its planning cycle in September while talks from GRC are still fresh.

By the time I was selected to chair the conference in 1992, the Medicinal Chemistry GRC faced challenges from heavy oversubscription, an overly strong orientation to industrial research, and a lack of diversity. So we worked hard to bring in more women, to invite students, and to keep a focus on unpublished research, including the work of academic scientists. Additional progress was made in subsequent years in bringing greater diversity to Medicinal Chemistry, which has strengthened the conference as well as the field in general.

There is a bright future for medicinal chemistry in the pharmaceutical industry. Drugs will always be needed. While we will continue to see advances in large molecules, proteins, and antibodies, our ability to pay for them is limited. In the future a balance will be struck in the development of large and small molecules. Despite current difficulties in the politics of the pharmaceutical industry, new medicines continue to be important. Medicinal chemistry will be the foundation for future pharmaceutical advances, and the Medicinal Chemistry Gordon Conference will continue to be the premier venue where scientists come together and share advances in the field, just as they have for the last seventy-five years.

James Bristol
Pfizer Inc.

Discoveries in
Medicine
& Friendship

Paul Anderson
Merck, retired

In 1997 the Medicinal Chemistry Gordon Conference celebrated its fiftieth anniversary at Colby-Sawyer College in New London, New Hampshire. From its inception to date the conference has been one of the most important annual gatherings of medicinal chemists from industry, academe, and government. It provides a forum to discuss the chemistry and biology of therapeutic substances as well as the technology used in their discovery. Open discussion and free exchange of ideas about medicinal chemistry are central to the spirit of the conference and are the driving force behind its heavy oversubscription year after year. The format of the Medicinal Chemistry Conference has always provided a unique opportunity for experienced investigators to encourage and transfer knowledge to the next generation of scientists in the field of drug discovery.

Insight into the modern history of drug discovery and its dramatic contributions to the high quality of health care we enjoy today can be found by reviewing the past programs for this conference. Topics have covered progress in antibiotic research; new medicines for controlling high blood pressure, preventing ulcers, and managing cholesterol levels; advances in cancer chemotherapy; the latest methodology for drug discovery; and many other timely subjects. The conference provided a platform for early discussion of seminal work leading to the discovery and development of histamine-H2 antagonists for ulcer prevention. The discovery of angiotensin-converting enzyme inhibitors was also presented here early in the evolution of this important treatment for high blood pressure. More recently the Medicinal Chemistry GRC has hosted important discussions about the discovery and use of HIV protease inhibitors and HIV reverse transcriptase inhibitors. Conference participants have included many scientists recognized as innovators of these and other important discoveries, including winners of the Lasker Award and the Nobel Prize. It can be said truly that this conference has always been poised at the frontiers of science.

Gordon Research Conferences were not intended to be all work and no play. For many years a Thursday-evening skit performed by members of the conference highlighted the lighter side of annual proceedings. A more recent tradition pits scientists from the East and West coasts in an annual softball challenge for the coveted Krantz Cup. These activities and many informal opportunities for social interaction have created a spirit of camaraderie among Medicinal Chemistry Conference participants on which many enduring friendships and professional relationships have been built.

Miming the
Genetic
CODE

The Gordon Research Conferences have supported important exchanges of findings about the genetic code and helped scientists working in the area learn new methods. In 1961 Heinrich Matthaei and I published the "poly-U experiment": we built a strand of messenger RNA (called the "poly-U strand") made only of the base uracil. Our experiment caused a bit of a sensation because it proved that ribonucleic acid acts as a genetic messenger. In the year that followed, we were busy deciphering the genetic code at the National Institutes of Health—and going to a lot of symposiums—when I attended my first Gordon Conference on nucleic acids.

Marshall Nirenberg
National Institutes of Health

Fritz Lipmann of the Rockefeller Institute, the greatest biochemist of his time and a magnificent scientist, chaired a session of the 1962 conference called "In Vitro Polypeptide Synthesis: Effects of Nucleic Acids" where I presented a paper. Joseph Speyer from Severo Ochoa's laboratory at New York University also gave a talk. Ochoa was one of our greatest competitors, but he had received me very graciously in his lab in New York at one time. Heinz Fraenkel-Conrat talked about the collaborative work we had done together on protein synthesis when I visited him for a month at the University of California at Berkeley.

I chaired a session in 1965 on polypeptide synthesis and another in 1967 on protein synthesis. When I moved into the field of neurobiology in the 1970s, I found that Gordon Conferences were critical meetings in this discipline as well.

I always thought the Gordon Conferences were among the best meetings of the year. You not only heard the latest things that were happening in the field, but you also had a chance to meet the participants and get to know them personally.

The informality of the sessions probably left the deepest impression on me. After all these years I still remember a "talk" Seymour Benzer gave in 1962 using two blackboards with pieces of words written on them. He selected these pieces just by pointing at them, and he gave his whole seminar—all fifteen minutes of it—without saying a single word. You might have thought he was trying to simulate the genetic code in mime (he was "talking" about RNA), but actually he was just trying to be funny. We were elucidated and entertained at the same time.

Perhaps this brief description of the Medicinal Chemistry Gordon Conference can best be summed up with the words of the conference anthem (sung to the tune of "Old Father Thames"):

High in the hills, down by the lake,
Happy and fancy free,
Medicinal folks keep rolling along
Back to the GRC.

What do they know, what will they learn,
A great deal of what is to be.
Medicinal folks keep coming back
Back to the GRC.

They never seem to worry,
Care not for sun or rain,
They never seem to hurry,
But they get there just the same.

Meetings may come, meetings may go,
Whatever their fortunes be.
Medicinal folks keep coming back
Back to the GRC.

Where the Cloning Discussion Began

Maxine Singer
Carnegie Institution
of Washington

Today understanding of DNA and RNA is fundamental to all biology. In the mid-1950s, however, the annual Gordon Conference was the only venue where those interested in DNA and RNA could meet, while still having to share the forum with protein chemists. Watson and Crick's double-helix model was still news, and for some it was still controversial. A few biochemists and renegade chemists and biologists were struggling to understand these large molecules. Money for science was scarce, as *Sputnik* had not yet been launched. Some biochemists and geneticists foresaw their fields merging into molecular biology, but the relationship was rocky, and the wedding date was indeterminate. Through most of the 1950s the annual Gordon Conference called Proteins and Nucleic Acids was at the center of this dynamic. The conference was held at the New Hampton School, where the countryside was beautiful, but the facilities (with their narrow, lumpy beds) seemed primitive, contrasting starkly with the meeting's revolutionary science and influx of new insights and colleagues each summer.

Geneticists appeared occasionally at the conferences: Alfred Hershey was on the program in 1954, the year after he and Martha Chase nailed down Oswald Avery's earlier proposal that genes are made of DNA. In 1956 there were initial reports on protein synthesis and the first descriptions of the enzymatic synthesis of polyribo- and polydeoxyribonucleotides by Severo Ochoa and Arthur Kornberg, respectively. Starting in 1957, proteins and nucleic acids were emphasized in alternate years; this was formalized in 1959 when GRC leadership deemed the scope of research in both areas "so wide" that each would have its own conference. Though still titled Proteins and Nucleic Acids, the first meeting

devoted entirely to nucleic acids took place in 1960. Finally, in 1962, a conference called simply Nucleic Acids was held, chaired by Leon Heppel and Cyrus Levinthal.

Throughout the 1960s precise new enzymatic tools for analysis of DNA and RNA and fresh fundamental understandings were described at the Gordon Conferences, including the biosynthesis of DNA and RNA, the semiconservative replication of DNA (the idea that replicated DNA molecules are composed of one original strand and one newly synthesized strand), the genetic code, the sequence of tRNAs (transfer RNAs), and the chemical nature of mutation. The general excitement forged lasting collegial relationships and friendships among those in a passionate and growing, but not yet enormous, community. Dieter Söll and I appreciated the great honor of being elected cochairs for the 1973 conference.

Daniel Nathan's report at the 1971 conference had unlocked vast opportunities for manipulating DNA in ways previously unimaginable. He showed us how, using the newly understood restriction endonuclease enzymes, the genome of SV-40 (and by implication all other genomes) could be reproducibly divided into smaller segments by cleaving its sugar-phosphate backbone. By the 1973 conference the entire Thursday-morning session was reserved for the discussion of results obtained with restriction enzymes. It was then that Herb Boyer electrified us by describing how foreign DNA segments could be cloned in *Escherichia coli* using a plasmid vector and pointed out the general applicability of this new method.

The story of what followed has been told often. Many in attendance that week began to imagine the enormous scientific potential of this new technique. Some raised questions about potential associated risks. John

As a molecular biologist I have throughout my career wanted to understand the nature of life. It has been my good fortune to live in the greatest period thus far in the history of biological science—the era when the machinery of life became evident, when enzymes and genes and viruses became tangible, when subcellular structures, fibers, particles, channels, and vesicles emerged discretely from the obscurity of protoplasm. The Gordon Research Conferences played an instrumental role in these advances. Their own internal genetic structure—formal sessions tempered by discussions during free time—allowed new ideas to combine and recombine in a way that greatly advanced molecular biology during the last fifty-five years.

**Robert L. Sinsheimer**
University of California at Santa Barbara

I came to molecular biology from an interest in math. As an undergraduate at the Massachusetts Institute of Technology (MIT) I switched from engineering to a new program in biological engineering. Since the department was so new, I took an eclectic range of courses. During World War II, I was involved in developing and testing airborne radar systems as a staff member, but after the war I returned to biology and completed my graduate studies at MIT. It was an exciting time because researchers such as George Beadle, Edward Tatum, and others were linking genes to enzymes, identifying DNA as the source of inheritance, and proving that nucleic acids could be genetic material.

When I finished my Ph.D., the job market for biophysicists was not strong, mostly because few people had heard of the field. Fortunately, Iowa State University had an opening in the physics department. Once there I was the only person in a large geographical radius working on nucleic acids. Toward the end of my first year at the university in 1949, I saw a notice that a Gordon Conference called Physical Methods in Nucleic Acid and Protein Research was to be held in the summer of 1950. Situated far from the action on either coast, I applied to attend and was accepted. The conference was a remarkable event: there were some ninety-five participants, including almost everyone who was doing physicochemical, structural, or metabolic studies on nucleic acid or protein molecules. In the years after the 1950 meeting the field expanded significantly.

(continued)

Sedat and Edward Ziff, two of the younger scientists at the conference, urged Söll and me to take time from the final scientific session on Friday morning to consider the implications of DNA cloning, or what came to be called "recombinant DNA" experiments. Because of time restraints the Friday discussion focused only on how to ensure that serious attention would be paid to the potential possible risks. Two actions were approved by the majority of those remaining (and eventually by a mail ballot, in order to include participants who had already left the conference on Friday). The first decision instructed Söll and me to send a letter to the presidents of the National Academy of Sciences and the Institute of Medicine describing the new method, summarizing the concerns that had been raised at the conference, and asking the academy to establish a study committee to recommend appropriate actions. The second decision was to publicize the letter more widely by submitting it for publication to the journals *Science* and *Proceedings of the National Academy of Sciences*.

Neither Söll nor I had foreseen that the opportunity to chair the conference would lead to new responsibilities for initiating the public discussion of this new science. The ensuing public and scientific debates, which have been documented in many books and articles, made the 1973 Nucleic Acids Gordon Conference famous. This event in history, however, marked what was to be only the beginning of the conference's enduring legacy as a major venue for the remarkable advances of biomedical research over the last thirty years.

During the first half of 1953 I took a six-month leave of absence to work with Max Delbrück at the California Institute of Technology (Caltech). This was a critical period: Linus Pauling gave a seminar describing a three-strand helix for DNA, and a short time later James Watson and Francis Crick came out with their two-strand helical model. My own research at the time shifted to phage genetics, specifically a focus on Phi X 174, a virus that can grow in *Escherichia coli*. In 1957 I was invited to set up a lab at Caltech. In the coming years we were able to identify key features of Phi X, including its ring structure as a novel, single-stranded form of DNA. We eventually synthesized an infective DNA strand in 1967 using DNA polymerase together with DNA ligase and Phi X DNA.

Throughout this period of research I actively sought out the most recent work in the field. Most good research would of course ultimately appear in publications, but only after a delay of six months or more. To hear it sooner, I became a regular attendee at the summer Gordon Conferences.

I was asked to chair the 1960 Proteins and Nucleic Acids Conference, the last year when both fields were discussed at the same GRC. By the early 1960s hundreds of applicants sought to attend, so we split the series into two conferences—one devoted to proteins and the other to nucleic acids. The Nucleic Acids Conference continued to grow and soon alternated each year's focus between structural studies and metabolic and functional studies.

It is a pity that better records were not made of the Gordon Conferences in the 1950s and 1960s. Here great advances and significant smaller steps were first presented, discussed, and criticized in the informal, open, and friendly fashion that GRC supports. Here Fred Sanger presented the first amino acid sequence of a protein—specifically, insulin. The sequencing of the adrenocorticotropic hormone was not far behind. At a Gordon Conference, Julius Marmur and Paul Doty described their remarkable discovery of DNA renaturation, which is the basis of much of today's molecular genetics. The enzymology of DNA and RNA was progressively elaborated. Here, too, Francis Crick presented his "adaptor" hypothesis to link nucleic acid sequence to amino acid sequence and his "commaless" mode. Mahlon Hoagland and Paul Zamencik described their discovery of transfer RNA. Bacterial transforming factors and viral structures were analyzed, and the relationship of DNA structures to genetic factors was progressively refined. Gradually, year by year, the modes of synthesis of DNA, RNA, and proteins were clarified.

We live in a unique time in the history of life on Earth, the time when a species first began to understand its origins, its inheritance, and its biological functions. Gordon Conferences have been there for every step of this remarkable journey.

My first moments at a Gordon Conference are permanently etched in my memory. I arrived at the Tilton School campus in Tilton, New Hampshire, in the late hours, about 11:00 p.m. I recall walking down dark stairs and suddenly finding myself in a bright room. It was extremely intimidating: I knew absolutely no one in the room; I was the only woman; and as the only graduate student, I was probably the youngest person attending the meeting by about ten years. Just as I was considering bolting back up the stairs like a lost Tilton high school student, one of the meeting organizers, Morton Bradbury, started talking to me about my science and made me feel incredibly welcome. As these few moments demonstrated, Gordon Conference participants are bound by science.

This was the 1984 Nuclear Proteins, Chromatin Structure, and Gene Regulation Conference, which had all the required elements of a Gordon Conference: cutting-edge science; endless hours of delightful speculation; discussion about a technique that was beaten to death (the footprint gel); a roommate (Sally Elgin); and finally, after the first or second late night in the bar, a stern sign that appeared by the pool prohibiting swimming after midnight.

I was fortunate to have been able to attend this GRC. My Ph.D. adviser, Tom Cech, could not attend and convinced speaker Joan Steitz to let me take his place in her session. Joan talked about small nuclear ribonucleoprotein particles (snRNPs) and, from what I remember, presented the first evidence that U4 and U6 (where U stands

Lessons Learned AND THE New Hampshire Police

for uridine-rich) snRNPs were in the same complex. I talked about ribozymes—catalytic RNA molecules. At this meeting I also met the late Hal Weintraub, who was to become my postdoctoral mentor. Hal presented his current work on enhancers (to which proteins bind and thereby increase the rate of gene transcription)—mostly in the form of a chalk talk. He also talked about experiments being done in his lab by John Izant using antisense RNA, a strand complementary to an mRNA, to alter gene expression. I was already seeing from a "ribo-centric" perspective at this point in my life, so these experiments got my attention. John Burch, a postdoc in Hal's lab, introduced me to Hal at the meeting, and a year later I began my postdoctoral work in Hal's lab.

My goal was to understand why the antisense technique worked well in *Xenopus* oocytes but not in *Xenopus* embryos. As part of a control experiment I injected a double-stranded RNA (dsRNA) molecule into an embryo, which set the stage for what is still the focus of my research. When I compared the dsRNA before and after injection into the embryo, I discovered something perplexing: on a denaturing gel the RNA strands migrated the same way before and after incubation in the embryo, but on a native gel the RNA had an altered mobility.

In 1988, when I arrived at the Nucleic Acids GRC, which I was attending every year by that time, conference organizer Norm Pace asked if I would accept a last-minute invitation to speak. Of course I said yes. My observations of the dsRNA's odd behavior captured the imagination of fellow attendees, and I remember the long hours of listening to everyone's ideas about what had happened. With the ideas and input of my fellow scientists I finally figured out that the RNA was covalently modified: I had stumbled on an RNA editing enzyme now known as an ADAR (adenosine deaminase that acts on RNA).

I cochaired with Dick Gumport the memorable 1994 Nucleic Acids GRC. This was GRC's last year using the New Hampton School site (in New Hampton, New Hampshire), which, although rustic, was beloved by the Nucleic Acids conferees; however, it was the conferees who helped put an end to GRC's use of the site. I received a phone call only days before the meeting, informing me

that we could not use the auditorium on Tuesday night because the headmaster had double-booked it. We were given other options, but in the end Dick and I decided to hold an afternoon rather than an evening session, thus violating the sanctity of reserving Gordon Conference afternoons for swimming, hiking, and discussing science. Further, the headmaster's group was scheduled to use the auditorium at a precise time following our session. Although we did not vocalize it, both Dick and I knew our session would probably not finish on time. Never in our wildest dreams did we imagine such a dramatic outcome of running overtime.

My only explanation for what ensued is that scientists like drama. Because GRC had reserved the site far in advance, many attendees felt enormously slighted about being asked to vacate the auditorium. When our session did indeed run over, we heard whispers and rumblings outside the auditorium, and the conferees went into battle mode. The headmaster's assistants started bringing me a disruptive stream of "warning notes" to end the session, instigating two disgruntled conferees, Dan Gottschling and Mark Roth, to barricade the doors. It was a surreal experience to be listening to elegant talks on structure with two conferees guarding the door like thugs. The headmaster continued to tug at the doors, and since the science was being compromised, I decided to attempt negotiation. I got up at the end of a talk, gave the secret handshake to the men guarding the doors, and left the auditorium to talk to . . . the New Hampton police! There I was, about to be arrested. I did not end the session immediately, but that was the end of New Hampton as a GRC site.

In 1997 another Gordon Conference came into my life—a conference devoted wholly to its namesake, RNA Editing, which began largely through the efforts of Harold Smith. Initially I was skeptical of a conference with such a narrow focus, but the meeting drew researchers from many other fields, and interest in the relationship of RNA editing to other fields grew. This was yet another lesson I learned from a Gordon Conference—this time without the police.

Brenda Bass
University of Utah and Howard Hughes Medical Institute

The Nobel Path

Paul D. Boyer
University of California
at Los Angeles

Spirited exchange of ideas and open discussion are critical to the advance of science. I have long felt that the American research university is one of the finest features of our society because it fosters creative cooperation among well-trained scholars. Assembling a group of faculty, postdoctoral fellows, and graduate students to discuss experiments creates a marvelous venue for the development of new knowledge. The Gordon Conferences operate in the same way; in fact, they offer a regular meeting site for leading minds to discuss novel ideas over the course of many years.

I entered scientific research in 1939 as a graduate student in biochemistry at the University of Wisconsin. At the time I took for granted the excellence of the department; I did not realize how fortunate I was to be able to walk down the hall from my lab and listen to the symposium of Nobel laureates Otto Meyerhof, Fritz Lipmann, Carl Cori, and others talking about respiratory enzymes and the discovery of oxidative phosphorylation. Later I came to see just how much work it takes to make good scientific dialogue possible.

My interest in protein structure and function first arose through research I did on human serum during a postdoc at Stanford University. After my postdoctoral year I began teaching at the University of Minnesota. Like others in my generation, I got a boost from the end of World War II. Graduate students were flocking to the universities, and new federal funding was available for molecular biology. After seventeen years in Minnesota I accepted an offer in 1965 to serve as the initial director of the Molecular Biology Institute at the University of California at Los Angeles.

That same year I attended my first Gordon Research Conference, called Energy Coupling Mechanisms. There are many competing meetings now, but for thirty years the Energy Coupling Mechanisms GRC was the premier way to communicate in the field. Every scientist doing pioneer work wanted to attend. In fact, the European Bioenergetics Conference series was started by GRC attendees and intentionally alternated years with the Energy Coupling Mechanisms GRC. Together, these meetings inspired a wave of breakthrough research in biology.

At my first GRC I presented results showing a previously unknown pathway for adenosine triphosphate (ATP) synthesis. Throughout nature, energy derived from sunlight or the use of oxygen is captured to make ATP, which is then used to drive many essential biological processes, such as muscle contraction, brain and nerve function, and the syntheses of protein and DNA. I mistakenly presented our results as the primary route for ATP formation. Some conference participants aptly questioned this interpretation, which stimulated our further studies showing that what we had found was a minor pathway for ATP synthesis. The major pathway remained tantalizingly unknown.

In the early 1970s we were still trying to find out how most ATP was made. Advances in bioenergetics showed that a complex enzyme called ATP synthase was involved. Earlier data led me to propose a new concept for ATP formation: that the major use of energy is not in the formation of ATP but rather in releasing tightly bound ATP that was readily formed at the catalytic site of the ATP synthase.

In 1970 I started participating in another GRC series called Enzymes, Coenzymes, and Metabolic Pathways. This conference was more general in scope but included sessions about ATP formation, and many of its attendees provided valuable appraisals of my research. Participation in both series played a vital role in our identification of the

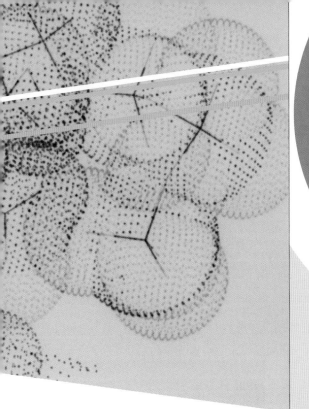

binding change mechanism for ATP synthesis, which involves three identical catalytic sites that participate sequentially in substrate binding, the formation of tightly bound ATP, and finally ATP release.

In 1981 findings in my laboratory and by others led me to propose that the binding changes of the catalytic sites were driven by the rotation of a centrally located subunit. We did not publish much on rotational catalysis at first because we had already communicated the idea to most of the important scientists in the field at Gordon Conferences. Through good fortune, at the 1981 Bioenergetics GRC, John Walker presented his preliminary studies on the X-ray structure of ATP synthase. Walker and I kept in touch, and it was gratifying when his structural results published in 1994 strongly supported the binding change mechanism with rotational catalysis. Walker and I shared half the 1997 Nobel Prize in chemistry for understanding of this mechanism.

With increasing specialization in science fewer opportunities arise to learn about important findings in allied fields. Although specialization is an inevitable result of growth in knowledge, it can become counterproductive by causing us to lose perspective on neighboring areas relevant to our work. Gordon Conferences present a solution to this side effect: they are structured in a way that promotes informal discussion alongside formal sessions and keeps us talking to our colleagues. In the future I hope GRC continues to hold fields together and broaden our research interests.

Feminism
IN TWENTIETH-CENTURY CANCER RESEARCH

At the suggestion of George Snell at the R. B. Jackson Memorial Lab in Bar Harbor, Maine, I attended the 1950 Cancer Gordon Research Conference. I was working on an American Cancer Society grant to show definitively that tumors could (or could not) evoke an autologous immune response. The conference was the start of a lifelong affection and admiration for GRC.

After serving as the first full-time program director for the extramural immunology area at the National Cancer Institute (NCI) of the National Institutes of Health (NIH) from 1970 to 1974, I chaired the first Tumor Immunology Gordon Conference in 1976. This conference was enthusiastically received by the scientific community and continued until 1979, when it ended because a chairman failed to send GRC headquarters the conference program. All was not lost, however, since various other Gordon Conferences—Cancer, Chemotherapy of Experimental and Clinical Cancer, Angiogenesis, and Proteins—address important aspects of tumor immunology.

Mary Fink
National Cancer Institute, retired

My experiences in cancer research have been both joyful and miserable. After marrying my husband (now of fifty-two years), I left Bar Harbor to take a position in the bacteriology department of the University of Colorado Medical School. I was hired at almost the same time as two men whose salaries were greater than mine but who lacked my experience in research and publication. I was, in fact, one of only two scientists at the medical school who held research grants. Although the medical students thought I was a good teacher, the department head thought my re-search interfered with my teaching. In spite of this I pursued confirmation that inbred mice could mount an immune response to their own tumors (as demanded by a reviewer of a paper submitted for publication while I was working at Jackson Labs), as evidenced by both the failure of a second transplant to grow and by an anaphylactic hypersensitivity reaction to a tumor extract. The reaction was exhibited in vivo and by the in vitro Schultz-Dale reaction. In preparing for the use of both techniques, I did a comparative study on various strains of inbred mice using egg albumin as the antigen. I found that histamine was not involved in the immune response but that serotonin very likely was. Years later, when I was no longer at the university, a paper was published by the microbiology department on mouse anaphylaxis without ever referring to our several previous publications on the subject.

After seven years I left the University of Colorado to work at NCI. My work there over twenty-five years was primarily on immune responses to leukemia and tumor viruses like Rous sarcoma and Epstein-Barr in mice, humans, and cats. These years were exciting and productive; the sometimes fourteen-hour days were like play. However, after being repeatedly rejected for promotion while male colleagues with far less scientific stature and experience were favored, I accepted the position of program director for immunology in the NCI extramural area.

(continued)

Immunology was an underdeveloped area at NCI, and my boss, the director for all cancer therapy research, had little time to devote to it. After developing a successful program in a few years, I became the associate director of biological research programs through a competitive selection process. The rewards of the job were few, however: two colleagues, one envious of my apparent popularity with grantees, the other jealous of my selection over him, overtly and covertly bad-mouthed me to a superior. My title was changed to "acting" associate director for three years until the position was permanently withdrawn with the official explanation that it never existed.

I spent my last few years at NCI as the special assistant to the director. Among the job's most interesting and rewarding aspects was serving as executive secretary to the National Cancer Advisory Board (NCAB), its Subcommittee on Special Actions, and the Agenda Subcommittee. "Special actions" are requests for board approval of innovative grants for work deemed not sufficiently meritorious by reviewers: the research of Baruch Blumberg, who won a Nobel Prize in physiology or medicine for his work on hepatitis, and the flourishing work of Judah Folkman on angiogenesis were once funded this way. When I retired from NCI after exactly twenty-five years, it was reassuring to be told that it took five staff members to replace me.

My experiences with GRC have always been enjoyable. After serving for six years as a member of the Selection and Scheduling (S&S) Committee, I served on the council for three. As I neared retirement, Alex Cruickshank invited me to run for the board of trustees. I became the first female trustee and represented the biological sciences for two terms. Not only did this experience compensate for the hateful discriminatory treatment I received at NCI, but it also wonderfully bridged my transition into retirement. Now eighty-five years old, I have so many wonderful moments to remember. I developed true admiration for chemists and physicists on the S&S committee and on the board. I learned a lot while monitoring conferences in various diverse biological subjects, and my husband, who is a psychologist, participated by revising the conference grading sheet and monitoring conferences of appropriate topics. Alex Cruickshank displayed extraordinary enthusiasm and generosity at Gordon Conferences and trustee meetings. The Gordon Research Conferences as they were then would not have existed but for Alex and Irene Cruickshank.

By providing as much time for discussion as for presentation, often about the unpublished results of ongoing research, the conference format created an air of pure scientific inquiry that is sadly lacking in many scientific conferences today. Sessions scheduled in the mornings and evenings left the afternoons free for play or individual discussions—an aspect of GRC that has since been copied by many groups.

I am cheered by the contrast in appreciation of female scientists today compared with that of the mid-twentieth century. This is especially visible in the makeup of the present GRC board. I was privileged to associate in my work with outstanding women (several of whom experienced similar negative treatment in their careers), including Sarah Stewart, discoverer of the polyoma virus, and Bernice Eddy, discoverer of the S-V40 virus; Yvonne Barr Balding (the "Barr" in Epstein-Barr virus); Ingegard Hellstrom, who pioneered studies in tumor immunology; Thelma Dunn, expert cancer pathologist; Virginia Evans and Kathryn Sanford, cell biologists and pioneers in tissue culture methodology; and members of NCAB such as biologist Janet Rowley and epidemiologist Maureen Henderson.

To an old cancer researcher it is heartening to learn that today aberrant stem cells may indeed be responsible for the etiology and metastasis in malignant disease. Perhaps the secrets of this scourge are at last being revealed and may lead to its being conquered.

My first experience with the Gordon Research Conferences came in 1965. I had just finished my residency in surgery at Massachusetts General Hospital and had set up a laboratory at Boston City Hospital to investigate the results of an experiment dating back to my two years in the early 1960s at the Naval Medical Research Institute in Bethesda, Maryland. I wanted to know why melanoma cancer cells became dormant in rabbit thyroid glands perfused with hemoglobin in isolated glass chambers but grew into large tumors and sprouted capillary networks in mice. This had remained a burning question for me, but few leading scientists were interested in cancer research at the time, since many considered it career dooming to pursue research into such a hopeless disease. But I read about the 1965 Cancer Gordon Conference in *Science* magazine and applied to attend.

I was enthralled to meet the great scientists there, like Charles Heidelberger, who was making tremendous progress in chemical carcinogenesis at the University of Wisconsin, and Harry Eagle, who had developed the first animal cell culture ten years earlier. I also met Richmond T. Prehn, with whom I would exchange ideas for the next twenty years. I was amazed that each speaker had the floor for a half-hour, unlike the mere ten minutes allotted at most other scientific and medical meetings.

I attended almost every Cancer GRC after 1965. For a number of years I did not tell anyone about the rabbit thyroid experiments because our conclusion that tumor growth depends on angiogenesis (the growth of new blood vessels) had been ridiculed by reviewers. But in

A Burning Question

1971 the *New England Journal of Medicine* published a talk I gave at Boston's Beth Israel Hospital describing the ideas that angiogenesis was stimulated by a growth factor and that antiangiogenesis agents could possibly serve as effective cancer therapies. I was then for the first time invited by Prehn to speak at the next Cancer GRC. My talk stimulated a heated debate. Many pathologists argued that tumors had no blood vessels: they were simply white masses of tissue. As a surgeon I had removed many tumors from the body and had observed that they were filled with blood vessels. But there were no other surgeons at the conference to back up my observations. And besides, other attendees argued, how could one distinguish blood vessels feeding tumors from the inflammation that tumors were thought inevitably to cause?

At first the theory that tumors were angiogenesis-dependent was by nature difficult to demonstrate: the endothelial cells that make blood vessels had never been grown outside the body, and imaging technology that could show angiogenesis at work inside the body did not exist. Arguments presented at the Cancer GRC motivated me to find a way to gather the necessary evidence. GRC also brought together scientists from different disciplines who eventually contributed enormously to the understanding of cancer and angiogenesis. I remember one meeting in particular at which polymer scientists helped explain why putting a piece of plastic in a mouse's skin causes a sarcoma. Rakesh K. Jain, who was originally a chemical engineer, helped biologists understand the fluid dynamics of vascular systems.

Of crucial importance for the field of angiogenesis research was the observation of Alexander Cruickshank,

then GRC's director, that interest in vascular biology and angiogenic inhibitors was growing at various conferences. At his suggestion Jain and I organized in 1995 the first Angiogenesis and Microcirculation Conference. Although in the early 1980s the medical community was still skeptical that angiogenesis inhibitors would be useful in patients, basic scientists had moved into the field from many different disciplines and from many countries. GRC provided a forum for the rapid advances taking place in understanding the molecular biology of the angiogenic process. Postdoctoral students and researchers from government and industry came to learn what was happening in the field, meet experts, and make funding and career connections. In fact, in the 1970s and 1980s Katherine Sanford and Colette Freeman from the National Cancer Institute gave me advice about grant writing because they heard me speak at a GRC.

GRC enabled the development of the field of angiogenesis research, which led to a modern understanding of cancer, macular degeneration, arthritis, and hemangiomatosis as angiogenesis-dependent diseases, among others. Between 1995 and 2002 the number of papers about angiogenesis increased almost sevenfold, with nearly 1,400 papers published in 2002. Since the 1971 paper in the *New England Journal of Medicine* appeared, more than 22,000 papers have been published with the word *angiogenesis* in the title. New and old medicines, such as endostatin, Avastin, TNP-470, and thalidomide, now show promising results as therapies that have begun to prolong survival in cancer patients. These developments were greatly facilitated by GRC's early recognition of the validity of scientific work in the field of angiogenesis research.

Judah Folkman
Harvard Medical School

Seminal Findings
and an Allergy to Black Flies

Rakesh K. Jain
Harvard Medical School
and Massachusetts
General Hospital

I am pleased and honored to contribute to this festschrift in celebration of the seventy-fifth anniversary of the Gordon Research Conferences. I have had the privilege of participating as a speaker, discussion leader, and organizer at more than twenty Gordon Research Conferences since 1990. These meetings have spanned areas as diverse as cancer, chemotherapy, radiation oncology, angiogenesis, vascular biology, microcirculation, drug delivery, and optics. They have allowed me to learn from the leaders in all these fields, to initiate new collaborations, and to present provocative ideas to the best brains in the business.

Although I had previously heard of these prestigious meetings and had seen their program announcements in *Science*, it was not until 1990 that I received my first invitation to speak at a Gordon Conference—on solute exchange in microcirculation and on drug delivery. I attended the Drug Carriers in Biology and Medicine Conference in New Hampshire in June. On the third day my mentor, Pietro Gullino, invited me to join him on a hike to the White Mountains during the afternoon break. We had a delightful time hiking in the hills and enjoying the scenery. By evening I was feeling a little tired but nevertheless prepared for the session, and I sat in the front row next to Pietro. After a short time I began to feel dizzy and excused myself to lie down. Soon I lost my vision and told Pietro I was feeling ill. The next thing I knew, I was in the hospital getting adrenaline. I had had an allergic reaction to being bitten by black flies. So my first lesson from the GRC was that I am allergic to insect bites. Since then I have limited myself to canoe trips.

Beyond this seminal finding I was quite struck by the caliber of the speakers, the cutting-edge nature of the work presented, the extent of formal and informal discussions, and the opportunity to mingle with the participants. I was equally struck by the fact that these two groups, microcirculation and drug delivery, were not taking advantage of one another's expertise, although they were addressing complementary questions. This lack of communication among different disciplines became even more vivid when I attended the Cancer Conference in 1991 and the Chemotherapy of Experimental and Clinical Cancer Conference in 1992.

In the majority of presentations cancer biologists viewed solid tumors—the cause of over 85 percent of cancer-related deaths—as a collection of cancer cells. However, from our own work and that of others, we knew that a solid tumor is made up of more than just cancer cells: it is composed of cancer cells, host cells, and blood vessels, all of which are embedded in an extracellular matrix. The tumor is an organ—albeit an abnormal one— but there was no GRC that treated it as such.

I saw an opportunity to combine these different areas—microcirculation, drug delivery, cancer, and chemotherapy—and proposed a new GRC called Angiogenesis and Microcirculation. I approached Judah Folkman with this idea and asked him to be my cochair. He was enthusiastic. We presented our proposal to Alexander Cruickshank and Carlyle Storm of the GRC, and our proposal was accepted. The first GRC on this topic was held in August 1995 at Salve Regina University, and these meetings have remained popular ever since: they continue to be oversubscribed! This is only one example of how the GRC has enabled scientists to develop new, interdisciplinary areas of study.

Becoming a Player
on the Field

The year 2006 marks not only the seventy-fifth anniversary of the Gordon Research Conferences but also the thirty-fifth anniversary of the Mammary Gland Biology (MGB) Conference. My first GRC experience was in 1973 when, as a second-year postdoctoral fellow, I was asked to give a plenary talk at the second biennial MGB Conference. I had entered the field only two months after the conference's first meeting, then called the Biology of Milk.

The MGB Conference has been a place where long-standing collaborations were formed and scientists grappled with fundamental issues in the field. The highly successful *Journal of Mammary Gland Biology and Neoplasia* was founded over lunchtime discussions, and its editorial board now meets annually at the conference. It was at this conference that Kurt Ebner explained the newly discovered function of the old milk protein alpha-lactalbumin. Dale Bauman described his progress in the use of recombinant bovine growth hormone to increase the milk yield of dairy cows (setting off an international firestorm, not to mention giving agribusiness a multimillion-dollar boost). When Dolly was cloned from a mammary epithelial cell, we spent an entire afternoon sitting on the floor of the parlor in one of the Colby-Sawyer College dorms debating the scientific, societal, and ethical implications of this earth-shaking event. The first transgenic pig, in which the mammary gland was used as a bioreactor to express blood-clotting proteins in milk through directed expression to the mammary gland, was first described at the 1987 MGB Conference, which I chaired.

Barbara K. Vonderhaar
National Cancer Institute

At the first conference I attended, I was delighted to meet and interact with the stars of the field—senior investigators such as Ken DeOme, Ranu Nandi, Stu Patton, Bob Jenness, Jim Linzel, Howard Bern, Joel Elias, Cliff Welch, and Dorothy Pitelka. We "smart young things" who wanted to be players would spend the morning and evening sessions listening to the "great white heads," as we called them. During the afternoons—on the tennis courts or while playing basketball, or late into the evening over drinks—we discussed the scientific insights they had shared with us. Today I am one of the "great white heads," and although golf is now my game, discussing new insights over evening GRC sessions or afternoon recreation remains as thrilling as ever.

The evolution of the MGB Conference, as well as the GRC organization as a whole, reflects the changes I have seen in science and the mammary biology field over the last thirty-five years. At the 1973 conference I was one of only a handful of women on the program or in the audience. In 1987 I was only the second woman to be elected chair of the MGB Conference. This field has since exploded: recognition of the importance of the

(continued)

In June 2001, at a Cancer Chemotherapy Gordon Conference, I proposed that antiangiogenic therapy not only prunes tumor vessels but also normalizes the tumor vasculature and makes it more efficient for drug delivery. This new and provocative concept received a mixed response from the audience, which galvanized my thinking even more. I went home the next week and wrote a commentary for *Nature Medicine*, outlining this hypothesis and its implications. I also began several new clinical and preclinical studies. The discussion at the GRC and the *Nature Medicine* commentary that ensued have served as guides for the current phase of my work. Emerging preclinical and clinical data from various laboratories now provide compelling evidence in support of the normalization hypothesis. Thus, the Gordon Conferences have been critical to my career development and growth and have provided a new direction for the field of angiogenesis.

In short, my own experience is a good example of how the seventy-five years of the GRC have allowed the fruitful association of areas of research and how the conferences have been the ideal forum for presenting new concepts. I believe the conferences are a benefit for all of us, and I look forward to their continuance in the years to come.

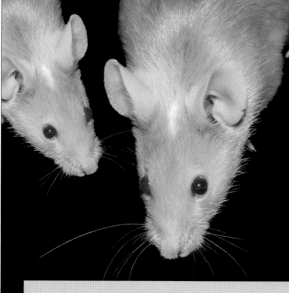

Becoming a Player *on the Field*

mammary gland as a biological system and as a source of food and energy, and the pervasiveness of breast cancer, have commanded extensive funding from public and private entities. Attendance at the MGB Conference steadily increased over the years until it became clear in 1999 that the community could support an annual meeting.

Because international attendance at this conference has always been high, the meeting now alternates between sites in the United States and in Europe. The conference is also vastly oversubscribed. Changes in demographics are equally important: today; at least 50 percent of the attendees at this conference are women, and five of the last ten chairs were female.

My sixteen-year involvement with GRC governance began in 1985 when I was first elected to a three-year term as an at-large member of the council. I then served six years on the Selection and Scheduling (S&S) Committee starting in 1989. I remember the first day of the 1989 S&S meeting at the York Harbor Inn in Maine, where once again I was only one of a few women in a sea of male faces. It was then that I first met Leila Diamond from the Wistar Institute, a legend in the cancer field. As the second woman to be elected to the board of trustees and an old hand at S&S, she took me under her wing. She and I became immediate friends and coconspirators. At the end of the fall 1990 S&S meeting, on one of our long walks along the rocky coast of Maine, she predicted that I would be elected to the board and one day serve as its chair.

I was only the fifth woman ever to be elected to the board, serving from 1995 to 2001, and only the second woman to become chair of the board, functioning as such for the 1998–1999 term. At our last meeting together in 1995, as I prepared to take her board seat, Leila and I talked about the culture of science and the lack of women scientists in prominent roles. We decided we would do something about it and that the GRC organization would be at the cutting edge of the issue. By the time I left the board, half of the twelve elected board members were women; since 2001 three more have served as board chairs. I believe that the Gordon Conferences and the entire scientific community have greatly benefited from this inclusion.

Over the past thirty-five years women have slowly advanced into the upper ranks of scientific society. The MGB Conference and the GRC organization as a whole have been at the cutting edge of this trend and stand as stellar examples of the inclusion of women in scientific pursuits for the rest of the community. I am proud to be associated with both.

The year was 1971, and Bob Jenness of the University of Minnesota and Stuart Patton of Pennsylvania State University teamed up to organize the Biology of Milk Gordon Research Conference. Beginning in the 1960s, the work of such "apostles" as Kenneth DeOme, Satyabrata Nandi, Howard Bern, Yale Topper, and Sefton Wellings and their protégés emerged to lead the field of mammary gland biology and breast cancer research. The conference would host a combination of breast cancer biologists and mammary physiologists who wanted to expand the field beyond the focus on nutrition and cancer studies of most breast studies of the period.

The first Biology of Milk Conference was a success and a revolution for the field. Several of the apostles of mammary gland biology were present. Of special interest was the identification of the milk protein alpha-lactalbumin as the B protein component of the lactose synthetase enzyme complex in the mammary gland (i.e., alpha-lactalbumin is a major protein of breast milk and is required for lactose synthesis). The conference continued to meet every two years and in 1977 was renamed Mammary Gland Biology (MGB). Throughout the 1970s the MGB Conference featured fascinating aspects of lactation biology, oncogenic activity, and in vitro three-dimensional growth systems. Pioneering these observations were the apostles, their protégés, and an international contingency from Holland.

The next decade saw the arrival of the transgenic mouse. Later in the decade Philip Leder led this area and gave an entertaining talk at the MGB Conference about MMTV-*ras* and *myc* transgenic mammary glands. About the same time Lothar Hennighausen and his collaborators talked about transgenic pigs that expressed human clotting proteins in the mammary gland and secreted them into milk. A new genre arose out of these presentations referred to as "signaling transduction in the mammary gland." Signaling concepts in the mammary gland would

JOURNEY OF THE MAMMARY GLAND
Apostles and Protégés

later propel mammary biologists into today's field of the molecular aspects of mammary gland development and cancer.

As a young, impressionable graduate fellow at the National Cancer Institute (NCI) in Bethesda, Maryland, I had just begun to hear the terms *fat pad* and *outgrowths* and learn about slow-release pellets and tissue-specific promoters that target the breast. Under the guidance of the second-generation mammary biologist David Salomon, I was exposed to historical perspectives on mammary gland biology and to its pioneers, all of whom attended the MGB Gordon Conference religiously. After hearing about the apostles and working alongside two of them—namely, Barbara Vonderhaar and Gilberth H. Smith—I was looking forward to my first MGB Conference.

I had previously attended several meetings about cancer but for the most part felt out of place, bored, and somewhat demoralized—probably because there were few young urban minorities like myself in attendance. My first experiences at GRC were totally different: I had a chance to meet, converse, jog, and play basketball with some of the mammary gland biology apostles. Today the environment of the MGB Gordon Conference environment still allows for the free exchange of ideas, information, and techniques; most important, it is free from barriers like egotism and elitism.

I remember how in the early 1990s Charles Daniel brilliantly explained the nuances of mammary ductal elongation, branching, terminal end bud formation and components, and stem-cell contributions to the mammary duct. Daniel's nomenclature, histology, and biological knowledge date back some forty years and form the foundation that most mammary gland biologists and breast cancer researchers use today. Other recent topics include Daniel Medina's oft-used cell culture–derived hyperplastic immortalized cell lines and their cytogenetic characteristics and Robert Cardiff's tutorial on the pathological aberrations or "new definitions" of breast cancer morphology. I should add that I will never forget Cardiff's mastery on the basketball court, or how, twice in ten years, he swept me in four consecutive games of twenty-one, even if he did cheat in one of them.

The MGB Conference gave me a chance to sit down and discuss retroviral fat-pad injections with Daniel and Smith, which led to fruitful collaborations. I also had the opportunity to give my first mini-symposium on the location of stem cells in the mouse mammary gland. I remember an E. F. Hutton–like moment while sitting at the lunch table with Robert Clarke, Elizabeth Anderson, Nicholas Zeps, Jeff Rosen, Jeff Green, and a few other covert ears nearby: they all asked, "How did you do it—double-label stem cells in vivo?" Using the data and discussions from this talk, most, if not all, mammary gland biologists attempting to label and transplant stem cells in vivo have used my techniques (5-bromo-2-deoxyuridine [BrdU]—a chemical base analog of thymidine used to detect the proliferation of living cells—and fluorescent cell tracers) since 1999.

Since first attending the MGB Conference, I have also witnessed a number of extremely talented women present research work ahead of its time, and I greatly respect their proactive quests to follow in the footsteps of Pitelka and lead us to the cure for breast cancer.

The conference still holds to its original mission—to provide a forum for the exchange of new knowledge, while being pleasant and entertaining. Much like a coach's relationship to an athlete, the MGB Conference has guided the direction of ongoing breast cancer research. As a history buff I have always found the MGB Conference refreshing and somewhat humbling in that some of the lecturers are finally seeing the fruits of their long journey. What a wonderful treat the MGB Gordon Conference has been.

Nicholas J. Kenney
Hampton University

Astonished and *Charmed*

Arieh Gertler
Hebrew University of Jerusalem

The late 1970s and early 1980s marked a major turning point in my scientific career. Up to that time my research had been devoted to the protein chemistry of proteolytic enzymes and their inhibitors, when I made a major decision to switch my focus to biological models of mammary glands and the hormones that affect growth and metabolism. The late Yale Topper from the National Institutes of Health, one of the pioneers of the field, helped me make that transition, partly by introducing me to the rich atmosphere of scientific openness inherent in Gordon Research Conferences.

The first GRC I attended was the 1979 Mammary Gland Biology Conference, and I was first astonished and then charmed by the informal atmosphere, with its open and free exchange of broad-ranging scientific ideas. I became a frequent attendee, as either a participant or a speaker, of the Mammary Gland Biology and Prolactin Gordon Conferences.

Many of my projects and collaborations with researchers from North America (such as Topper, Bob Collier, John Byatt, Craig Baumrucker, Bob Bremel, Nira Ben-Jonathan, and Li Yu-Lee) and Europe (Paul Kelly, Jean Djiane, and Richard Vernon), as well as many friendships, emerged from informal talks and discussions at those conferences over a bottle of beer or a glass of wine or during breaks between tennis sessions. The resultant interaction with Collier had a particular impact on my research. It culminated in a six-month sabbatical at the Monsanto Company, which opened a window for me onto the world of large-scale recombinant protein production and subsequently led me into site-directed mutagenesis studies of several protein hormones.

The Mammary Gland Biology Conferences are unique because they bring together researchers from three different disciplines. Each group studies this complex and unique gland from a different vantage point: the molecular and cellular biologists work on the basic processes in the tissue; the researchers study basic and clinical aspects of breast cancer; and the agricultural scientists are interested in the milk-production processes. So the conference is very productive and mutually rewarding. Similarly, the focus of the Prolactin Gordon Conference slowly evolved into a broader topic encompassing all hormones in the mammary family, including growth hormones and placental lactogens.

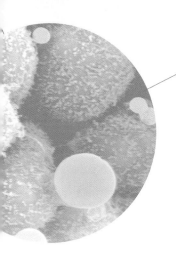

Because no official records are kept of what transpires at the Gordon Conferences, no one can quantify their contribution in terms of numbers of papers or citations. However, I am sure that I express the feelings of many when I say that GRC should be considered a major scientific tool, not only because it contributes to many international collaborations and friendships but also because its structure facilitates the emergence of new ideas and concepts.

My studies in chemistry began in 1955 at the University of Sheffield in the United Kingdom. Sheffield was my hometown, but only later in life did I recognize the university's excellent reputation for chemical and biochemical research. My studies there stimulated a lifelong fascination with biomolecules, how they are produced, their structures, and their modes of action.

In 1961 I began postdoctoral studies with David Sprinson at Columbia University, where every staff member of the biochemistry department was a textbook name. I learned a lot there and loved New York. I became interested in biological electron transfer processes, which were a very hot topic in 1963, and I gained experience in this field through further postdoctoral studies at the universities of Oxford and London. My mentor in London, where I studied the metallo-enzyme xanthine oxidase, was Bob Bray. At the end of my stay in his lab he kindly allowed me to work independently on the nature of copper in pig plasma amine oxidase, recently purified by Hugh Blaschko and Franca Buffoni at Oxford. I collaborated with Buffoni on electron paramagnetic resonance spectroscopy studies that showed that the copper was in the oxidized (cupric) state and remained so following reduction with substrate amine. This surprising observation had also been made by Bruno Mondovi and Helmut Beinert in their studies of pig kidney amine oxidase. The role of copper in the catalytic mechanism of amine oxidases remains controversial to this day.

Peter Knowles
University of Leeds

In 1967 I took a lectureship in the biophysics department at the University of Leeds. My parents were relieved that I had a real job after so many years as a student. I continued work on pig plasma amine oxidase but was interested in other copper enzymes, particularly the cytochrome c oxidase (the membrane-bound enzyme that controls the last step of food oxidation), which seemed to hold the key to understanding oxidative phosphorylation. I knew very little about membranes, and so in 1972 I arranged a short sabbatical to visit the University of Virginia laboratory of Ching Hsien Huang. We had met when we were both postdocs with Manfred Eigen at the Göttingen Max Planck-Institut. During this visit several of us left from Charlottesville and drove up to Meriden, New Hampshire, to attend the Lipid Metabolism Gordon Conference, which was focused specifically on membranes that year. This experience, and my visit with Huang, gave me the confidence to start experiments with model membrane systems. These experiments have continued throughout my career and led to recent work on peptide self-assembly and bio-nanotechnology.

During the 1970s and 1980s my lab focused on analyzing the structure and catalytic mechanisms of amine oxidases using magnetic resonance and kinetic methods. After six

(continued)

years of detailed detective work we had established that the copper centers had two coordinated water molecules, both of which were important to catalytic function. Progress in those days was slow. To understand more about the structure of these proteins, we turned to protein crystallography. Since amine oxidases were big (180,000 daltons), we started with a smaller, related enzyme, galactose oxidase, whose beautiful structure we reported in 1991. In 1995 we reported the first structure of an amine oxidase, which has provided a sound foundation for continued work on the molecular basis of catalysis.

In the 1970s there was a growing awareness that the organic cofactor of amine oxidase was not pyridoxal phosphate but rather was a redox cofactor. The field was greatly stimulated in 1984 when it was reported that the cofactor was a quinine—pyrroloquinoline quinone (PQQ)—that had earlier been found as a dissociable cofactor in bacterial dehydrogenases. Further studies, particularly by Judith Klinman, revealed that the quinone cofactor (TPQ) was the quinone form of trihydroxyphenylalanine derived by redox modification of an active site tyrosine.

By 1990 a substantial population of researchers around the world was interested in quinone cofactors, and Paul Gallop and Herb Kagan organized a GRC called Quinoproteins and Pyrroloquinoline Quinone (PQQ). I returned to the Gordon Conferences after nearly twenty years.

It soon became evident that quinone cofactors like TPQ are only one type of protein modification. We now recognize a large family of enzymes with protein-derived radical cofactors as well as novel cross-links: galactose oxidase is an example with its cysteine-tyrosine cross-link and tyrosine radical cofactor. The title of this GRC has since evolved to encompass radical enzymes involved in rearrangement reactions as well as redox reactions: it is now called Protein-Derived Cofactors, Radicals, and Quinones. Study of these enzymes spans theoretical physics to medicine and involves physicists, chemists, and biological scientists.

My life has been enriched by the friendships I have built over the years attending the Protein-Derived Cofactors series and other Gordon Conferences. Even though we have all been driven by the rivalry of frontier science, we have completely trusted each other with regard to unpublished data. After chairing the Protein-Derived Cofactors Conference in 1997, I was nominated to the GRC board of trustees in 2000, in part, I believe, because I had perspective about the GRC from outside the United States. It has been wonderful to watch the GRC expand from New England to California and Europe, without financial loss and without losing sight of fundamental GRC-brand principles.

I am particularly excited by a new venture to establish GRC sites in developing countries, possibly starting with India. My own research lab has benefited from visiting Indian scientists—in particular, Kapil Yadav from the University of Gorakhpur, who while visiting in the 1980s and 1990s produced the first crystals of galactose oxidase and an amine oxidase, breakthroughs that were crucial to solving the structures of these enzymes. As Shobo Bhattacharya, the director of the Tata Institute for Fundamental Research in Bombay, said in a recent *New Scientist* article, India needs to present its best science in a context of the frontiers of world science, as should other developing countries. GRC offers the ideal framework for doing this by giving young scientists from around the world a chance to experience the stimulation of Gordon Conferences firsthand.

The Gordon Research Conferences have dominated my scientific career for more than thirty years. My first GRC venture came in 1972, when as a young research scientist at the Institute for Cancer Research in Philadelphia, I was invited by Gene Cordes to attend the Enzymes, Coenzymes, and Metabolic Pathways GRC. I presented my work on kinetic isotope effects in the alcohol dehydrogenase reaction, the oxidation of alcohol to aldehyde. I remember my terror in speaking for the first time in front of the brilliant "mafia" of enzymologists that normally lined the front rows of the lecture hall. This venerable conference still serves as a proving ground for young scientists and is an important venue for the interplay of chemistry and biology (and was so even long before the now-fashionable term *chemical biology* had been coined). The Enzymes Conference has remained an intellectual powerhouse, with many of the field's giants of the second half of the twentieth century participating, like Frank Westheimer, Mo Cleland, Ernie Rose, Bill Jencks, and Bob Abeles.

Other Gordon Conferences have also played large roles in my scientific career. Beginning in the late 1980s international interest was growing in the idea that certain new enzymatic cofactors were covalently linked to proteins, particularly that one low-molecular-weight bacterial cofactor called pyrroloquinoline quinone (PQQ) was attached to a large number of eukaryotic proteins. This attention instigated a new Gordon Conference in 1990 that focused on PQQ and was called Quinoproteins and Pyrroloquinoline Quinone. My laboratory had been struggling with one of the prototypic PQQ enzymes, a bovine serum amine oxidase (BSAO). In the year preceding the second GRC on PQQ we had finally sorted out the identity of the cofactor in BSAO, finding that it was indeed a

The MAFIA and the Melting Pot

quinone but one with a structure very different from PQQ. I remember experiencing the thrill of presenting data that were finally able to specify the nature of one of the new eukaryotic cofactor structures, but my feelings were mixed with some regret that I had failed to provide preprints of our work to the PQQ proponents in advance of the meeting. By the end of the GRC it was clear that the conference needed a name change. With the growing roles for novel cofactors and free radicals in enzymatic reactions, this relatively young field is now represented by a GRC called Protein Derived Cofactors, Radicals, and Quinones.

Another Gordon Conference has also played a dominant role in my research career—Isotopes in Biological and Chemical Sciences, a descendant of the Isotopes Conference (which is one of the oldest Gordon Conferences, having celebrated its fiftieth anniversary in 2004). The Isotopes Conference emerged in the aftermath of World War II because of the important role isotopes had played in weapons development. From the outset the conference was focused on the origins of equilibrium and kinetic isotope effects and their application to reaction mechanisms. A major figure in the field is Jake Bigeleisen, whose theoretical treatise on semiclassical isotope effects dominated the field for three decades.

The Isotopes GRC has always represented a unique melting pot of disciplines, welcoming physiologists, geochemists, and atmospheric chemists, for whom isotopes play a major role in research. In the 1970s enzymologists recognized the capacity of isotope effects to expand greatly the interpretive power of enzyme kinetics, and a new era was under way. Early on, data emerging from enzyme studies of hydrogen transfer showed discrepancies that appeared to be incompatible with the dominant semiclassical theory. My lab's experimental work ultimately played a key role in these revelations, and I remember well the initial resistance to the shifting paradigm. In the best tradition of the GRC, and illustrating the crucial interplay among disciplines in advancing science, biochemists and physical chemists with a common interest in isotopes moved toward theory dominated by quantum mechanical tunneling in hydrogen transfer in condense phase. As indicated by the most recent Gordon Conference on isotopes, Isotopes in Biological and Chemical Sciences, the implications of this tunneling behavior for our understanding of the origins of hydrogen isotope effects and enzyme catalysis continue to be a major source of debate and inquiry.

During my scientific career I have been privileged to participate in and serve as chair for each of the conferences discussed above. Some of the insights gleaned from these disciplines have spilled over into other conferences, such as the Metals in Biology GRC. For example, after hearing a lecture by Joe Berry from the Carnegie Institute on oxygen isotope effects in the O_2 side reaction of the enzyme rubisco, my lab initiated a series of experiments in which oxygen isotope effects serve as an important probe into the nature of O_2 activation in a diverse family of metallo-enzymes.

During the last three decades the guiding hands of Alexander Cruickshank and, later, Carl Storm have been evident throughout the progress of GRC. I am especially appreciative of Carl's vision and commitment to the ever-changing nature of science and, as a consequence, the conferences that represent the individual disciplines.

Judith P. Klinman
University of California at Berkeley

Cogs, KNOBS, &Switches

Dagmar Ringe
Brandeis University

My very first Gordon Conference was the 1982 Enzymes, Coenzymes, and Metabolic Pathways Conference at Kimball Union Academy in Meriden, New Hampshire. Together with Gregory Petsko I brought along an electron density map on stacked sections so that the enzymologists could see for themselves what the structure of an enzyme looks like and how crystallographers were getting their results. Seeing the arrangement of atoms in space could clarify how the parts come together to make such an exquisite machine, complete with cogs, knobs, and switches that could perform a chemical job unique to the work of an enzyme. There was only one problem with this engineer's view of the world; my favorite biochemist, Robert Abeles, told me that the structures we were seeing could not possibly be right because the results suggested rigid molecules—and enzymes are not rigid. This was the start of a wonderful, exciting, and rewarding journey through structural enzymology.

The work of our research group and of other structural enzymologists showed that enzymes are indeed not rigid and that they respond to the presence of ligands. Enzymes also harbor surprises that only their structure can show. In some cases even the details of how the cogs, knobs, and switches work can be demonstrated, which stimulates notions about how to interfere with them and thereby manipulate enzyme function.

Since that first conference in 1982 I have attended a number of different Gordon Conferences, some highly focused, others more general in scope. At each one I have learned a great deal, have met wonderful scientists, and have come away with new ideas and collaborations. In fact, one of my most fruitful collaborations, with Jack Kirsch, was initiated by a casual encounter at the swimming hole during the Enzymes, Coenzymes, and Metabolic Pathways Gordon Conference. I cannot imagine a better way of meeting—on a sweltering day, luxuriating in a cold stream and discussing the chemistry of biological systems. Other collaborations were born during tennis games or on hikes. I am still developing collaborations from such serendipitous meetings. These special encounters—the opportunities to talk about science in an informal setting—are what set the Gordon Conferences apart in my mind.

Infatuation with Decadence and Science

Science does not stand still, and new ideas are constantly changing the way we approach it. Structural biology has had a major impact on the way biologists think. The Gordon Conferences have played a major role in providing a venue where new ideas and new approaches can be learned and accepted. I have been fortunate to be part of the revolution that brought considerations of structure into biochemistry and biology, and I anticipate continuing to be part of the changes that new insights will bring.

Gloria C. Ferreira
University of South Florida

> Reflections

" When biotech was especially hot during the 1980s, the hotels for some Agricultural Science Conferences were filled with non-registered hangers-on trying to learn of possible investment opportunities. "

Elliot Bergman,
Griffin Corporation, retired

My infatuation with the Gordon Research Conferences began in 1989 when I was a postdoctoral fellow in Pete Pedersen's laboratory at the Johns Hopkins University School of Medicine. David Garboczi (now at the National Institutes of Health [NIH] but then also a member of Pedersen's lab) and I had decided to leave Sunday morning from Fell's Point in Baltimore and drive to the Holderness School in New Hampshire to attend the Molecular and Cellular Bioenergetics GRC. The plan was to arrive before dinner and the evening session. As the number of traveled miles and the number of ingested rice crispy cookies (which constituted our diet for the day's trip) increased, the idea of dinner became more and more appealing. Fortunately, we made it on time. There we were, having dinner and talking with the "names" we only knew from the literature! The opportunity to listen to and converse with Har Gobind Khorana, Robert Huber, John E. Walker, and Paul D. Boyer, to name a few, was an incredible treat for someone just trying to figure out if she would fit into the world of science.

Admittedly, I had the more immediate goal of learning the decisive news on membrane proteins and how to tackle their overproduction (or simply at that time their production) in the bacterium *Escherichia coli* and in *Saccharomyces cerevisiae* (otherwise known as baker's yeast). Clearly, if we were to get to the heart of the matter of transporters and enzymes, we would need material to work with, and the solution appeared to be the overproduction of recombinant proteins.

After the session on Sunday evening (a relatively early night) and back in my dormitory, I remember marveling at the decadence of the site: trickling shower water, more cold than warm; toilets separated by mere curtains; a mattress with clearly identifiable springs; and a single light hanging from the ceiling of the room I shared with another young postdoctoral fellow. Somehow it took me back to my days of camping as a child, which delighted me, and I felt inexplicably comfortable in this new environment and certain it was going to be a great week.

(continued)

Infatuation with Decadence and Science

And a great week it was. The lectures were wonderful, as were the questions, the heated debates, and the poster sessions that followed. I was trying to take in as much as possible, but more than the knowledge and information, it was the atmosphere in which they were transmitted that I found astounding. The accessibility of renowned scientists amazed me; the intensity of some of the arguments and discussions, combined with the ability to laugh, surprised me; and the long hours that extended into the night of talking about science or philosophy, exchanging lab protocols, and venting about our own last frustrating experiments made me realize how similarly investigators operate, how the achievement of goals unites them, and how universal science is.

I have attended several other Gordon Conferences—such as Chemistry and Biology of Tetrapyrroles and Enzymes, Coenzymes, and Metabolic Pathways—and they have all lived up to my first GRC experience. Their modus operandi remains the same and so has my enthusiasm for them. Now an independent investigator, I am a regular participant in the Tetrapyrroles GRC, and I am always excited to learn about new developments in the field, see old friends, and watch with pleasure the burgeoning of new scientists.

Through nearly two decades we have gone from being able to isolate and purify the enzymes involved in the synthesis and degradation of heme, to cloning of the genes encoding these enzymes and the determination of their three-dimensional structures, to developing model therapy systems for porphyries and identifying the role of tetrapyrroles as signaling molecules in various cellular processes. Gordon Conferences have facilitated progress in this field by fostering an atmosphere conducive to the exchange of ideas between established and new investigators from a wide variety of scientific disciplines.

I owe to Pete Pedersen my initiation into the GRC experience, to the GRC my growth as a scientist, and to the Chemistry and Biology of Tetrapyrroles Conference, in particular, the respect of my colleagues as a scientist.

> ## Reflections

" During my participation in the Basement Membranes GRC, I obtained new knowledge that I was able to share with my colleagues; this was very important for the developing science in Uzbekistan. Libraries in Uzbekistan today almost never receive foreign literature, so GRC is very important for scholars from my country. "

Bahrom Aliev
Institute of Virology, Uzbekistan

A First Love

A first Gordon Conference is like a first love—something you do not forget. The first Gordon Conference I attended was the Lipid Metabolism Conference in 1958, when I was a postdoctoral fellow in the laboratory of Konrad Bloch at Harvard University. Up until that time my experience with scientific meetings had been the spring gatherings of the experimental biologists in Atlantic City, a tawdry venue where thousands of scientists descended to listen to a week of ten-minute presentations by graduate students. It was a fine place to meet old colleagues strolling on the boardwalk, but students had little contact with the titans of their fields.

The Gordon Conference opened a new world for me. It marked a time when the study of lipids was emerging from *Schmierchemie* into an exciting scientific discipline. Realization had finally set in that biosynthesis was not simply a massive reversal of degradation, and such topics as fatty acids, sterols, and phospholipids had become hot areas of research. Most of the leading workers in these areas were present at the Lipid Metabolism Conference. We of the Bloch laboratory were excited about our recent discovery of branched alcohol pyrophosphates as intermediates in cholesterol synthesis, but we did not want to disclose this in the presence of our chief rival, Feodor Lynen. While hiking with Lynen in the afternoon, I made a comment about "interesting phosphate esters." Someone in the group said, "I heard that they were pyrophosphates. I visited Tchen in Boston and he told me that." Our colleague T. T. Tchen, who was not at the conference, had

already spilled the beans. Lynen smiled enigmatically, and I have no doubt that he was already neck and neck with us, or perhaps ahead. Sixteen years later Bloch and Lynen shared a Nobel Prize for their contributions to sterol and fatty acid synthesis.

Fatty acid synthesis was an important topic at this meeting. David Green reported on progress in his laboratory made by Salih Wakil, then a postdoctoral fellow. Acetyl coenzyme A (CoA) could be converted to longer fatty acids by liver extracts, but only if bicarbonate was present in the reaction mixture. At this meeting the idea arose that acetyl CoA might be converted to malonyl CoA by adding carbon dioxide, thus providing a better substrate for acylation with an additional molecule of acetyl CoA. Progress rapidly followed from this idea, and several laboratories soon established the important role of malonyl CoA in fatty acid synthesis.

Many lively Lipid Metabolism Gordon Conferences followed. The 1958 conference, however, represented a turning point in the field, and since it was my first GRC, it made an indelible mark on me.

John H. Law
University of Georgia

Fifty and Fifty

Edward A. Dennis
University of California
at San Diego

I remember fondly the first time I attended a Gordon Research Conference: it was the Lipid Metabolism Conference in the summer of 1969. At the time I was working as a postdoctoral fellow with Eugene Kennedy, who had discovered the main pathways for the biosynthesis of phospholipids. The department of biological chemistry at Harvard Medical School had many faculty and students who were interested in various aspects of lipid metabolism and frequently attended this conference. Many were then or later became leaders in the field.

The Lipid Metabolism GRC first met as the Blood Gordon Conference at Kimball Union Academy in 1955. The conference actually emphasized lipids more than it did blood, so it was renamed Lipid Metabolism in 1956. It has just reached its fiftieth anniversary. Interestingly enough, Kimball Union Academy was first used as a conference site in 1954, so the academy has also just had its fiftieth anniversary as a GRC site. The academy site, however, has since been replaced by the Waterville Valley Resort. Having attended Gordon Conferences at Kimball Union numerous times over the past thirty-five years, I will miss the country store that marks Meriden, New Hampshire; the beautiful views of rolling green hills; and the swimming hole by the covered bridge. But I will not miss the hard prep-school mattresses; the spartan, hot rooms; or the communal bathrooms.

The Lipid Metabolism GRC was always attended by the leaders in the lipid field. Nobelist Konrad Bloch and Eugene Kennedy were regular attendees. A lectureship named jointly after these two close friends and vanguards of the lipid field was established at the 2005 conference, for which the inaugural lecture was given jointly by Nobel laureates Michael Brown and Joseph Goldstein, both of whom have participated frequently in the Lipid Metabolism GRC.

I have vivid memories of the earliest conferences I attended, where vigorous debates occurred on subjects ranging from the validity of Arrhenius plots to the mechanisms involved in fatty acid synthase. On one occasion Salih Wakil, the biochemistry chair at the Baylor School of Medicine, and Roy Vagelos, then biochemistry chair at Washington University School of Medicine, had a shouting match about the role of acyl carrier protein in the synthase. The young turks of that time—Bill Lennarz, Armand Falco, and C. Fred Fox—urged them on. In those days the meetings were held in a hot, stuffy gym where participants had the choice of sitting on uncomfortable folding chairs or just outside the entrance door on the porch where you could hardly see the ancient black-and-white slides. Fortunately, by the 1980s Kimball Union had built a new air-conditioned conference room.

Back in *Fashion*

My research career began as a microbiologist, after DNA structure and gene function were discovered. During the first part of my career in the 1970s I witnessed the pioneering studies on recombinant DNA techniques and sequencing and the consequential possibilities of invaluable biotechnological and medical applications. By the 1980s the common view held that all the fundamentals of biology had already been discovered. As it was becoming easier to work with eukaryotes, it was our duty to move on to higher organisms to elucidate more complicated questions such as cell development and cancer. Moreover, the prevailing belief was that infectious diseases were under control owing to antibiotics. Microbiology therefore became a less fashionable and poorly funded discipline.

Nevertheless, some of us persisted in studying microbiology. This small community met at such Gordon Conferences as Bacterial Cell Surfaces, the Biology of Host–Parasite Interactions, and Mechanisms of Membrane Transport. The meetings generally brought out paradoxical results, passionate discussions, the exchange of advice and technical tools, and the establishment of new collaborations. We learned from skilled scientists how to choose and orient our own research themes. It was also an opportunity to appreciate recent scientific advancements. As the years went by, there were key discoveries in protein secretion, chaperones, transmembrane signaling, quorum sensing, sporulation, motility, chemotaxis, membrane protein structure, bacterial pathogenesis, drug resistance, and many other areas.

Cecile Wandersman
Institute Pasteur

Meanwhile, infectious diseases were again recognized as a major threat to public health, owing to the emergence of new diseases associated with AIDS and the materialization of multiresistant forms of bacteria caused by the uncontrolled use of antibiotics in the food and health sectors. This situation stimulated the scientific community to search for new antimicrobial compounds. Simultaneously, the avalanche of genome sequences provided us with mountains of data yet to be interpreted. As a result bacterial genetics remains one of the most powerful approaches to discovering new functions.

Today we understand that biology in general is much more complex than we used to believe, and we expect new models and theories to arise sooner or later. GRC has played a major role in keeping microbiological discussions running in the background, and I have no doubt that the Gordon Research Conferences will resonate with the future battles of biology.

In 1988 I had the pleasure of serving as vice chair and in 1989 as chair of the Lipid Metabolism GRC. Topic emphasis alternated between lipoproteins in the even years and lipid enzymology in the odd years, but the conference was separated in 1997, with Lipoprotein Metabolism meeting in the even years and Molecular and Cellular Biology of Lipids meeting in the odd years.

In the past decade it has been widely recognized that in addition to their traditional roles in membrane structure and energy metabolism, lipids and lipoproteins also play important roles in cell signaling and in most diseases. As a result both conferences should continue well into the future to become two of the longest-standing GRC meetings. As always, I look forward to presenting the latest work of my research group at the next GRC.

THE MEETING FOR "BONEHEADS"

Felix Bronner
University of Connecticut
Health Center

The first conference on metabolic interrelations, sponsored by the Josiah Macy, Jr., Foundation, took place in February 1949 in New York. Frank Fremont-Smith, the foundation's medical director, had realized that "the continued isolation of the several branches of science one from another is a serious obstacle to scientific progress" and that a multidisciplinary approach is particularly important for medicine. From the inception of the first conference calcium and bone were emphasized; by the fifth and last conference the topics covered dealt exclusively with the chemistry, physiology, pathology, and metabolism of bone and bone-related substances, and "with special reference to calcium" was rightly added to its title. As might be expected, the conferees were largely senior professors and investigators who often referred to work done by students, fellows, and junior colleagues.

In 1953, at the International Congress of Physiology, where I gave my first paper on the calcium metabolism of adolescent boys (which had also been the subject of my doctoral thesis at the Massachusetts Institute of Technology), I met William Neuman and Harold Copp, both of whom were my seniors. All three of us bemoaned the end of the Macy Conferences. I suggested the organization of a Gordon Conference on the subject and proposed that young investigators should also be able to participate. I was charged with writing to the American Association for the Advancement of Science to propose the conference.

My letter remained unanswered until near Christmastime, when I received a telephone call from GRC director George Parks, asking me to chair a conference on calcium. At that time I was still a postdoctoral fellow at MIT with Robert Harris, who had been my thesis adviser. Surprised but delighted, I asked Bill Neuman and Reidar Sognnaes (at that time at Harvard Dental School, and later dental dean at the University of California at Los Angeles) to join me in forming an organizing committee. I chaired the first Chemistry, Physiology, and Structure of Bones and Teeth Gordon Conference in 1954, with Neuman and Sognnaes as cochairmen (Copp was in Vancouver, Canada, and felt he was too far away to assist in the organization), and Neuman, Sognnaes, and Copp chaired the conference in 1955, 1956, and 1957, respectively.

From the beginning the Bones and Teeth Gordon Conference was an important scientific event: at that time it was the only meeting point for "boneheads" that was distinguished by cutting-edge reports. Moreover, the still-standing GRC policy that forbids publication of the meeting contributed significantly to the openness and collegiality of the conference. For thirty-two years, from 1954 to 1986, the Bones and Teeth Conference met annually at Kimball Union Academy in Meriden, New Hampshire, with maximum attendance (by mostly academics, with industry attendees constituting a small minority). Conferees have come away with ideas for further work, collaborations, and friendships. The conference has also stimulated the formation of many societies and other conferences, including the European and Israeli symposia on calcified tissues; conferences on parathyroid hormone, calcitonin, and vitamin D; and such societies as the American Society for Bone and Mineral Research, the International Bone and Mineral Society, the European Calcium Society, among others. Despite the various opportunities worldwide for meeting with colleagues and sharing knowledge on bone and calcified tissues, the Bones and Teeth Gordon Conference (called Cell and Molecular Biology of Bones and Teeth from 1997 to 2001) continues to meet in alternate years, attracting researchers from around the globe.

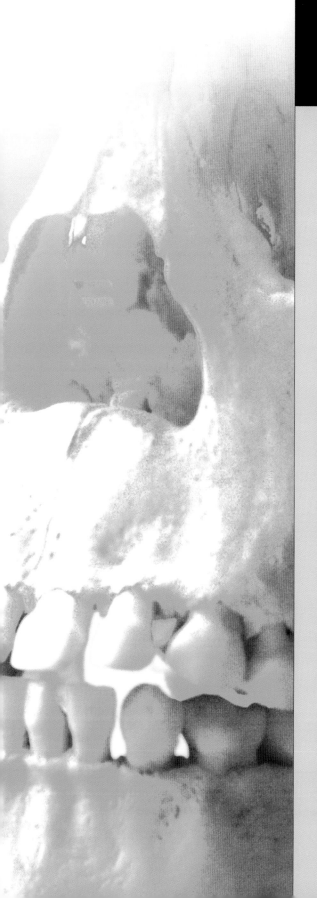

EQUAL PARTS *of* PLANTS, FUNGI, &
METAZOA

The field of epigenetics addresses the control mechanisms superimposed on primary DNA sequences. When the Epigenetics Gordon Research Conference was first proposed in 1994, the importance of epigenetic effects in gene regulation was only beginning to be appreciated. The first conference, organized by Timothy Bestor and myself, was held at the Holderness School in Plymouth, New Hampshire, in 1995.

Three people working on epigenetic effects on gene expression—Robert Martienssen, Amar Klar, and Bestor—contacted me in 1994 asking if I would be willing to help organize a Gordon Conference on epigenetics. There had previously been only one open meeting on this subject—a Keystone Symposium, organized by Shirley Tilghman and held in October 1993, which emphasized genomic imprinting in mice and humans. A few small meetings brought together people working on different systems who shared an interest in epigenetics. We all felt that the time was ripe for a regular, open meeting and that the format of GRC would be perfect. The GRC proposal that Tim and I wrote was approved, with enthusiastic support from Nina Fedoroff, Joachim Messing, Martienssen, and Richard Jorgensen, who were all working with plants; Jasper Rine, Eric Selker, and Jean-Luc Rossignol, who were working with fungi; and Tilghman, Davor Solter, and Steven Henikoff, who were working with animals.

Vicki Chandler
University of Arizona

Ninety people attended the first meeting; there were twenty-nine speakers, and about fifty posters were presented. We met our goal of bringing together scientists who had related interests but little opportunity to interact by including equal numbers of scientists who studied plants, fungi, and metazoa. The participants were very enthusiastic about the meeting and voted to hold another one in 1997. The participants also voted that there would be two cochairs per conference, each of whom worked with a different organism, to continue broad representation at future meetings.

For the past ten years organizers of the Epigenetics Gordon Conference have kept with the themes established at the first meeting: successfully bringing together scientists studying epigenetics in diverse organisms like mammals, plants, fungi, insects, protozoa, and prokaryotes; focusing the meeting on genetic and developmental consequences of epigenetic effects; exploring molecular similarities among seemingly diverse phenomena; and encouraging the presentation of findings that would be deemed "weird" at most other meetings. It is clear that many seemingly diverse phenomena in

(continued)

different species share common underlying mechanisms. At the 2005 meeting speakers were instructed by the organizers, Chao-Ting Wu and Judith Bender, to push the envelope and speculate with no holds barred about the implications of their research, resulting in the most stimulating, thought-provoking Epigenetics GRC to date.

Looking back at the program for the first conference is illuminating. Major emphasis was placed on gene silencing in plants, animals, and humans. Conferees were describing the phenomenology and looking for common themes but at that point had little understanding on a molecular level. In the years since, the molecular mechanisms underlying the phenomenological topics have begun to emerge, and other conference topics reveal more mechanistic understandings.

Changes in the field are revealed by examining what was not on the radar screen ten years ago. A dramatic example is the discovery that RNAi (RNA interference) in the nematode worm *Caenorhabditis elegans* and quelling (gene silencing in fungi) in molds of the genus *Neurospora* are mechanistically equivalent to post-transcriptional gene silencing and co-suppression in plants. RNA-mediated silencing featured prominently in subsequent meetings. Another example is that in 1995 we knew about histone modifications, but the major focus was on histone acetylation and deacetylation. Research on histone methylation, which received little attention in the early years, has exploded.

The Epigenetics GRC draws on many fields. Jean Finnegan, the 1997 cochair, relayed to me a memory of a dinner at the first conference where Klar, whose area of expertise was yeast genetics, asked everyone at the table, as well as the puzzled wait staff, whether they and their family members were left- or right-handed. Fascination with the genetic control of handedness has matured over the years, and at the 2005 meeting Klar presented the hypothesis that segregation of differentially marked DNA strands could contribute to handedness and certain human diseases.

Rob Martienssen, who cochaired the 2001 meeting, told me about how Rich Jorgensen made an offhand remark in his talk at the same conference that post-transcriptional gene silencing in plants seemed to occur on polysomes (polyribosomes, or several ribosomes configured to simultaneously translate the same messenger RNA). Rob wondered whether the argonaute proteins might be involved in gene silencing. That evening Andy Fire gave a talk on RNAi in *C. elegans*. Just prior to that, Craig Mello had called to tell him the identity of *rde1*, a gene required for RNAi, so Andy wrote it on a slide that he showed in his talk. It was argonaute! Rob had been working with the *ago1* (argonaute) gene in plants for some time, and more recently in *Schizosaccharomyces pombe* (fission yeast). He grabbed Andy to show him pictures of plants and yeast: it was a moment Rob will not forget.

The Epigenetics GRC is a truly multidisciplinary conference. Epigenetics has become a field unto itself by the gathering at this meeting of scientists who have been able to compare and contrast their epigenetics research for a variety of organisms.

My first experience with the Gordon Research Conferences was in New Hampshire during the summer of 1970. I was close to the end of my postdoctoral training in molecular biology in Bernard Horecker's laboratory at the Albert Einstein College of Medicine, and I was ready to go back to Siena, Italy, to work on the mechanism of male hormone action on muscle metabolism. Attending the Hormone Action Gordon Conference had been exciting and unforgettable. Besides its scientific benefit, this event made me realize how useful Gordon Conferences are to newcomers by helping them get to, and stay at, the frontiers of science. In 1988, a year after I had shifted my

> Reflections

" The inception of the Plant Molecular Biology Gordon Conference really founded a new community of researchers. It was the first time we knew what to call ourselves. It brought together young and old; Europeans and Americans; people working on microscopic plants and those working on higher plants; biologists, biochemists, and plant physiologists; and DNA-RNA experts and geneticists working on plant systems. "

Mary-Dell Chilton
Syngenta

GETTING BETTER
with Age

scientific specialty to biogerontology, I recalled these benefits and decided to attend the Biology of Aging Conference in Ventura, California, as well as almost every one of its subsequent meetings.

I used the GRC format model when I helped organize a binational Italy-U.S. conference called Protein Metabolism in Aging, held in April 1989 in San Miniato (Pisa), Italy. Harold Segal and Morton Rothstein, of the State University of New York, Buffalo, and Piero Ipata, of the University of Pisa, were co-organizers. The conference was a success: it was likened to a Gordon Conference and was the right event at the right time.

James Florini, a biogerontologist and a GRC trustee, was a member of the American delegation at the binational meeting, and he thought San Miniato would be an inexpensive, appropriate site for a small- to medium-sized Gordon Conference. He proposed the idea to Alexander Cruickshank, then the director of GRC. A few weeks later Robert Grasselli, who was traveling through Europe searching for good GRC sites, paid me a visit in Pisa and toured the San Miniato site. An additional Italian site was also chosen—Volterra, a small, beautiful city nearby, rich in Etruscan and medieval monuments. A GRC delegation led by Cruickshank and Florini soon came and approved both sites. In September 1990 two successful Gordon Conferences were held in Volterra, and then two more in 1991 in San Miniato, one of which was Biology of Aging, chaired by Arlan Richardson. Substantial financial support had been obtained from local banks (Cassa di Risparmio di Volterra and Cassa di Risparmio di San Miniato), the Italian National Research Council, and the Serono Foundation, which made the conferences in Italy as inexpensive and convenient as those in New Hampshire and California.

A committee for GRC European activities was established, with Robert Grasselli as chair and myself as the European vice chair. This was an unforgettable experience. I was able to serve GRC by providing guidance on local resources essential to the conferences in Italy. Other Gordon Conferences hosting hundreds of renowned scientists of many nationalities have since met in Italy; Il Ciocco, which is close to Pisa, is now a major GRC site in Europe. Italian science, especially biogerontology research and education, has benefited significantly from this: in the 1990s three Biology of Aging Gordon Conferences were run in Tuscany, and several international organizations as well as a European Union–sponsored summer school on biogerontology were started.

In the past decade the growing interest in the mechanisms of cell maintenance (including apoptosis and more recently autophagy) has been noticed by the GRC directors and board. This has been of crucial importance for the field of aging and age-associated diseases. In the last ten years new Gordon Conferences provided a forum for rapid advances in research in molecular and cellular biology of programmed cell death as well as cell organelle degradation and renewal. These were big steps forward to a modern understanding of aging, anti-aging intervention, and age-associated diseases, including neurodegeneration, atherosclerosis, and cancer.

Ettore Bergamini
University of Pisa

CONNECTING EAST TO WEST

Nancy Y. Ip
Hong Kong University of
Science and Technology

My first exposure to a Gordon Research Conference occurred early in my career, but its effect on my professional development has been tremendous and long lasting. In the early 1980s, during the course of my scientific education and training in the United States, I attended the Synaptic Plasticity Gordon conference. I was immediately impressed by its unique format. Researchers from all over the world engaged in intellectual exchange. At the same time focus on specific topics helped me assess the relevance of my research direction to the field of neuroscience as a whole. The opportunity to converse with and present my ideas to leading figures in the field invigorated me, and the ensuing discussions enhanced my research interest in neurotrophic factors and neuronal signal transduction.

The impact of GRC extends beyond personal development, however. Gordon Conferences have helped promote and enhance regional scientific excellence, especially scientific progress in Asia. After spending a brief period working in the biopharmaceutical industry in the United States, I returned to Hong Kong in 1993 to assume a faculty position at the newly established Hong Kong University of Science and Technology (HKUST). Since that time Asia has undergone astounding growth in the quality, quantity, and importance of its contributions at the forefront of science. In particular, many Asian countries have invested heavily in neuroscience research; the establishment of RIKEN Brain Science Institute in Japan and the Institute of Neuroscience at the Chinese Academy of Sciences in Shanghai, China, are examples of this investment in an institutional form. Hong Kong also made great strides in establishing a world-class research infrastructure and has emerged as a leader in molecular neuroscience.

In 1998 the region's neuroscience community received a major boost when the first Molecular and Cellular Neurobiology GRC was held in Beijing, China. The conference attracted scientists from across the region and overseas. I was subsequently elected vice chair of the conference's next meeting, scheduled to be held at HKUST in 2000, and was heavily involved in its organization. The 2000 conference was well received, and when the GRC board of trustees decided to establish a permanent Asian site at HKUST, we felt a huge sense of achievement. Since then two successful Molecular and Cellular Neurobiology Gordon Conferences have been held at that location—one in 2002, which I chaired, and one in 2004 for which I acted as the local coordinator.

Establishing a venue for Gordon Conferences in Asia has been an important step toward recognizing Asia's contribution to neuroscience. Asia's involvement in the field is confirmed by the participation in GRC of large numbers of scientists from overseas, as well as prominent researchers, including, for example, Nobel laureates Paul Greengard and Robert Horvitz, both of whom presented their lead findings during the 2002 Molecular and Cellular Neurobiology GRC. Horvitz also presented his work at the 2004 conference. Holding such a prestigious conference in Asia has also provided many Asian scientists, unable to travel to conferences overseas for various reasons, the priceless opportunity to engage with their international peers in their own backyard. It has also made Gordon Conferences more accessible to Asian students, thereby offering Asia's budding scientists invaluable exposure similar to what I received as a young scientist in the United States.

In its short history in Asia, GRC has already led to successful exchanges of research ideas, collaborations, and the development of competitive postdoctoral training opportunities for Asian students and other junior scientists. Furthermore, the intensive scientific interactions have advanced molecular neuroscience research and neurodrug discovery in Asia. In terms of my own research program, for example, I had the opportunity during the 2002 conference to discuss scientific interests with John Trojanowski and Virginia Lee from the University of Pennsylvania. Since these early encounters we have established collaborations focused on the discovery and development of small-molecule drugs isolated from traditional Chinese medicines that possess promising biological activities for the treatment of various neuropathologies.

By encouraging open communication about important issues at the frontiers of neuroscience, GRC plays an important role in the development and recognition of scientific excellence in Asia. GRC offers international forums for discussing and sharing the latest advancements and observations, and provides an environment for cultivating collaborative interactions. The significance of such an open dialogue should not be underestimated: it can facilitate the development of a truly integrated and collaborative global scientific community.

Learning One's Place
in the Universe

Eduardo Marbán
Johns Hopkins University

My first Gordon Conference, in 1980 in Tilton, New Hampshire, shaped my subsequent scientific life. I attended the Ion Channels GRC as a combination M.D. and Ph.D. student working in Dick Tsien's lab at Yale. My first paper had been accepted by *Nature*, and Dick had asked me to present our work. I thought I was pretty hot stuff. Other scientific conferences I had previously attended were formal and noninteractive, leading to little in the way of scientific street smarts.

GRC, however, was humbling. Many people present were smarter than I (sometimes much smarter), their life passions were consumed by focused scientific questions, and they were not shy about offering their opinions. I came away from this meeting with a new sense of my place in the universe: I was smaller, but because I had also learned a lot, I was larger in other ways. I went on to chair my own GRC called Cardiac Regulatory Mechanisms in 1998, but nothing could match the excitement of my first time.

From *Materia Medica* *to* Molecules

Palmer Taylor
University of California at San Diego

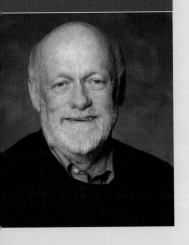

The first Molecular Pharmacology Gordon Conference was held in 1969, a time when receptors were known to exist only by inference from the selectivity of drug action, but no receptor had yet been chemically characterized. Moreover, little was known about the molecular basis of drug action, and in many cases one could not discern whether the target of drug action was a discrete receptor, ion channel, regulatory protein, particular DNA sequence, or some less selective site in membrane lipids or subcellular organelles involved in DNA replication, protein synthesis, and processing.

Receptors, or "receptive substances," were proposed by Paul Ehrlich and John Newport Langley around the turn of the last century; yet for seventy years receptors were defined by properties of function and recognition, leaving their chemical characterizations unresolved. In fact, the description of receptors as discrete chemical entities is arguably the cornerstone discovery of molecular pharmacology, and the emergence of molecular-based approaches coincided with seminal discoveries regarding receptor molecules.

In the late 1960s and early 1970s three developments contributed substantially to the discipline of molecular pharmacology. First, Avram Goldstein recognized the importance of communicating molecular-based research in the field of pharmacology and started the journal *Molecular Pharmacology*. Second, Arnold Burgen developed a Medical Research Council Unit in Cambridge, England, directed solely toward research on the molecular basis of drug action. And third, the first Gordon Conference devoted to this subject was developed. Each event contributed in different ways to the generation and dissemination of the molecular foundations of drug action. A look at the early conference chairs, topic organizers, and speakers shows continual participation of members of the Cambridge group and the journal's editorial board in these conferences.

Looking back at the literature of that period helps us understand how far we have come over the last thirty-five years. Many of the original concepts regarding the targets of drug action might now be thought of as primitive, but without the techniques to purify membrane proteins and characterize targets in low abundance, we had to develop working hypotheses to advance the field. Hence, proposals that acetylcholinesterase was a cholinergic receptor, adenylate cyclase was the beta-adrenergic receptor, and opiate receptors were proteolipids were subjects of papers at that time. These very subjects were debated with much spirit at the early Gordon Conferences. In each two-year cycle of the Molecular Pharmacology meeting one could see more complete characterizations of receptors and other drug targets, new ways to examine conformation of ligands and then soluble proteins that bound pharmacologic agents, and identification of the nucleic acids and lipids that were potential targets of drug action.

As was evident in the early years, advances in the discipline relied heavily on the gifts of nature: rhodopsin as a structural model for G-protein-coupled receptors; the electroplax of fish in the genera *Torpedo* and *Electrophorus* for obtaining nicotinic receptors and voltage-gated channels; candidate antineoplastic agents that bind to nuclear sites to block cell proliferation; and the peptide and alkaloid toxins that have proven so useful in characterizing drug targets. Such natural probes with a selectivity honed by evolution, when matched with their target identification, have also contributed heavily to understanding signaling and structure in modern biology.

Many new technologies have also enabled the field to emerge, such as nuclear magnetic resonance, fluorescence and other forms of spectroscopy, mass spectrometry, and novel affinity methods for purification of resistant proteins. Concepts underlying the recognition and signal transduction properties of drugs could be rigorously tested with these new technologies. Perusal of the list of speakers at early Gordon Conferences shows the continuing involvement of the pioneers in the field, many of whom carried major research endeavors over several decades and up to the current day.

Part of the continuing success of the conference came from participants in several of the sessions who were not aligned to the discipline of pharmacology. They brought novel technologies and systems that widened the research perspectives of the attendees. In turn, the conference brought new applications and research endeavors of therapeutic significance to those outside the pharmacological fields.

Comparing the ontogeny of the modifier *molecular* as applied to pharmacology and biology is of interest. *Molecular* in pharmacology applies to the working molecules of pharmacology: both the drugs and their structural relationships to neurotransmitters, hormones, and autacoids and the targets of drug action (primarily proteins, but secondarily nucleic acids or lipid sites). By contrast, the *molecular* modifier in biology became aligned with recombinant DNA technology. Hence, in the contemporary "-omics" vocabulary, molecular pharmacology has primarily a proteomic and secondarily a genomic base, whereas molecular biology has a primarily genomic base with proteomics being secondary.

Those of us attracted to this emerging field and who attended the early conferences at the New Hampshire sites reserved a portion of the summer for GRC. Between my interest in molecular pharmacology and my wife's interest in cyclic nucleotides and protein phosphorylation, Gordon Conferences meant traveling to the East Coast, parking our children with their grandparents in upstate New York, and coordinating our vacation time between the New England conference and the family cottage in the Adirondacks. Both were a welcome respite from an increasingly urbanized California.

SIGNAL TRANSDUCTION & DUNCE AWARDS

The Cyclic Nucleotides Gordon Conference played a major role in shaping my career and also helped to define a new area of knowledge that has come to be a central dogma of signal transduction. When I first attended the conference in the early 1970s, I was a young assistant professor who had been trained as a protein chemist but who chose to work on a newly discovered protein: cyclic adenosine monophosphate (cAMP)–dependent protein kinase (PKA). My first Gordon Conference in 1972 was actually Enzymes, Coenzymes, and Metabolic Pathways, which was the primary Gordon Conference for enzymologists, protein chemists, and crystallographers. It was my natural scientific home. But shortly thereafter my interest drifted to PKA, and I decided that I needed to learn more about this new area.

At that time the G-proteins, which bind the guanine nucleotides, were just being discovered, and my first Cyclic Nucleotides Gordon Conference in 1974 was chaired by Al Gilman. Here I met lifetime friends Shmuel Shatiel, Ed Krebs, Eddie Fischer, Jackie Corbin, Al Gilman, Henry Bourne, Ora Rosen, and Bruce Kemp, to name only a few. And here I got to know the pioneers of the field and became acquainted with pharmacology. In retrospect I recognize how much this Gordon Conference helped forge a bridge in my career between protein chemistry, structural biology, and pharmacology. The conference became my intellectual home for many years.

Susan Taylor
University of California at San Diego

The evolution of the name of the Cyclic Nucleotides Gordon Conference reflects the evolution of the field. Intellectually, the conference began with the discovery of cAMP by Earl Sutherland, who chaired the first conference in 1970. The discovery of G-proteins was revealed during the initial years of the Gordon Conference. The 1970s also saw the discovery of new kinases in addition to phosphorylase kinase, PKA, and the protein phosphatases that became part of the signaling equation. In the next decade the mitogen-activated protein (MAP) kinases and their cascades were discovered, and we learned that the insulin receptor was also a protein kinase. Eventually, the name of the conference was changed to Cyclic Nucleotides and Protein Phosphorylation.

Two branches of this Gordon Conference evolved, with the focus in one year on G-proteins and phosphodiesterases and in the alternate year on protein kinases and protein phosphorylation. The present title, Phosphorylation and G-Protein Mediated Signaling Networks, recognizes the two major facets of signal transduction: the G-protein-coupled receptors and signaling by protein phosphorylation.

(continued)

SIGNAL TRANSDUCTION & DUNCE AWARDS

My own research was influenced profoundly by the talks I heard in the early conferences and by the collaborations that resulted from afternoon discussions as we hiked and canoed. I chaired the conference in 1986 and attended regularly. At these conferences I heard about the latest discoveries. I remember most clearly the plenary lecture given by the late Ora Rosen where she talked about the insulin receptor and the differentiation of N1H3T3 cells into adipocytes. Her embracing of tissue culture made me also appreciate how important it is to learn new techniques that are driven by the science.

I have tried to follow this rule in my own career. When I first talked about trying to crystallize PKA, I was not taken very seriously by the GRC participants. In many ways I think I introduced structural biology into protein phosphorylation. I have been gratified to see how far we have come. I believe that my first talk about the protein kinase structure was at the Gordon Conference in 1991. Now, in this genomic era, we know that this is one of the largest gene families in mammalian genomes and over sixty protein kinase structures are available. Structural biology is now an integral and routine part of signal transduction.

One of the features that always made this Gordon Conference special is its interdisciplinary nature. It is a place where the disciplines of physiology, pharmacology, and biochemistry come together. It is also a meeting that specially fosters interactions with students and postdoctoral fellows.

Another highlight of this Gordon Conference is the "Dunce Awards." I think they were initiated early on by Larry Brunton and Lee Witters. Tom Sturgill and Bruce Kemp became later contributors. For as long as I can remember, this was a much-anticipated event that took place in the bar after the evening plenary lecture. A "secret" committee worked diligently to present three or four Dunce Awards each year, complete with cone hats for the recipients. One noteworthy example defines the flavor of these awards. Roger Tsien received the award one year for "the blush of conception"—his demonstration, using FRET (fluorescence resonance energy transfer), of the intracellular wave of cAMP that was initiated by the fertilization of an egg by a sperm.

Spontaneity always characterized the scheduled and unscheduled events at this Gordon Conference, and the combination created a unique environment. It brought together the conference attendees scientifically and socially and certainly facilitated communication and advances in the field.

> Reflections

"The Plant Molecular Biology Gordon Conference strongly encourages student and postdoctoral participation, and it is my impression that they enjoy the one-on-one discussions with leading scientists as much as those following each talk. I will always remember one of my students expounding on the cutting-edge science on his return and then adding, 'And I was shaving next to Meyerowitz!'"

Pamela J. Green
University of Delaware

Physical

Visions

Electron Spectroscopy:
Many Applications
for Many Scientific Frontiers

Cedric J. Powell
National Institute of
Standards and Technology

In 1974 an X-Ray Photoelectron Spectroscopy (XPS) Gordon Research Conference was held at Brewster Academy in Wolfeboro, New Hampshire, to discuss the interpretation of results from atomic physics, solid-state physics, and inorganic, organic, and analytical chemistry. The XPS technique, also known as electron spectroscopy for chemical analysis (ESCA), evolved from instrumentation and the publication in the 1950s and 1960s of numerous results from Kai Siegbahn's group at the University of Uppsala. (Siegbahn was awarded a Nobel Prize in physics in 1981 for his XPS accomplishments.) Commercial XPS instruments became available in 1969, and the 1974 GRC was a timely opportunity to discuss interpretational issues associated with the emerging technique.

The 1974 GRC on XPS evolved into Electron Spectroscopy Gordon Conferences that were held from 1976 to 1998 at two-year intervals. Although more than half of the 1976 conference was devoted to XPS topics, new topics were also included, ranging from photoelectron microscopy to low-energy electron diffraction, Auger-electron spectroscopy, and extended X-ray absorption fine structure. Presenters also discussed industrial applications of XPS and the use of XPS to identify unwanted surface species and reactions on pollutant particles in the atmosphere.

In subsequent years results from newly emerging electron spectroscopy techniques were discussed. These techniques made available such new information as the atomic structure in the outermost atomic layers of surfaces; the bonding configurations of atoms and molecules at surfaces and interfaces; surface chemical reactions; electronic excitations and dynamics in atoms, molecules, ions, solids, and liquids; high-resolution electron microscopy and microanalysis; and characterizations of atomic and molecular clusters. The presentations and discussions in these areas included the recent scientific advances and applications in catalysis, corrosion, microelectronics, magnetism, growth and characterization of thin films, high-temperature superconductors, and what would now be termed nanotechnology. Issues of interest were the fundamental properties and processes of ideal materials (e.g., atoms, molecules, and single-crystal surfaces) and similar properties and processes on materials of practical interest.

The discussions at the Electron Spectroscopy Gordon Conferences were extremely stimulating. Attending the conference were leading atomic and molecular scientists and solid-state and surface scientists who did not ordinarily attend each other's meetings. There was a mix of theorists and experimentalists with backgrounds in physics, chemistry, and materials science. I benefited greatly by learning about concepts and applications outside my background and by getting new ideas for my own research.

I served as a member of the GRC board of trustees from 1982 to 1986 and as chairman from 1985 to 1986. My term as chairman coincided with the transition of the GRC office from paper records for scientists applying and registering to attend a Gordon Conference to a computerized system. Not surprisingly, problems and delays occurred with this transition. This period also coincided with an extended sick leave for Alex Cruickshank, and I was naturally worried about whether registration and other arrangements for the 1986 summer conferences would be made on time. The GRC staff worked extremely hard during this difficult period, and the 1986 conferences were held successfully.

From Hot Atoms
to the
Stratosphere

In the summer of 1952 I received my Ph.D. in radiochemistry from the University of Chicago, working with Willard Libby. Newly married, I went off that fall to Princeton's department of chemistry as an instructor. The following summer I attended the Nuclear Chemistry Gordon Conference at the New Hampton School in New Hampton, New Hampshire.

F. Sherwood Rowland
University of California at Irvine

The physical sciences had undergone an enormous change during the 1940s, the product of spectacular changes in government funding during World War II. Nuclear chemistry had grown from a few scientists working with natural radioactivity to the use of ever-larger particle accelerators, nuclear reactors, and elaborate measuring equipment, much of it related to the development of nuclear weapons.

During the course of the wartime Manhattan Project a variety of new chemical specialties had arisen. In the early postwar period the Nuclear Chemistry Gordon Conference covered a broad spectrum of topics, including transuranium elements, radiation chemistry, new chemical separation techniques, age dating based on uranium and Libby's carbon-14, and my own niche, "hot atom" chemistry, which examines the unusual chemical reactions of radioisotopes as they emerge newly created with enormous energies on a chemical scale. As an example, tritium—the hydrogen isotope of mass three (hydrogen-3)—is created when a free neutron reacts with the nucleus of the stable isotope lithium-6 and is released with an energy of 2,700,000 electron volts. The triton (the singly charged nucleus without an electron) careens through its environment, leaving a long trail of molecular destruction. Eventually it picks up an electron and continues to move through the environment as an energetic neutral tritium atom. Only at the very end of its path does its energy decrease to a level roughly comparable to the strength of a carbon-hydrogen bond—4 to 5 electron volts—and undergo a final bond-forming collision. Because the tritium atom is radioactive, it can still be detected at the end of this trail whatever its terminal molecular form. Frequently, with very high kinetic energy in this last collision, such radioactive atoms end up in molecular forms inaccessible to thermal hydrogen atoms.

(continued)

After participating for several years in the annual meetings of the Selection and Scheduling Committee to review assessments of the previous year's conferences, I wondered whether the amount of available financial support for each conference affected the conference quality, as judged by the attendees and the GRC monitors. Although some financial support was essential to ensure participation of key scientists and discussion leaders, I found that there was, in fact, no clear correlation between the available measures of conference quality and total financial support. Assessments of quality are based primarily on the quality of the scientific presentations and discussions, the mix of attendees, and particularly the leadership and management skills of the chair. Apart from an orientation meeting, only rudimentary formal guidance was given at that time to new chairs, and I was motivated to prepare more specific information and expectations for speakers, discussion leaders, and overall management. This information, with further details, is now available on the GRC Web site.

I have subsequently organized about a dozen conferences in a Gordon-type format. Invited presentations were followed by substantive discussion periods, and a poster session allowed presentations by attendees. These conferences were both stimulating and rated highly by attendees, results that I attribute to my previous experience as a GRC chair and conference attendee. My own research has also profited from new ideas arising from formal and informal interactions at Gordon and Gordon-type conferences.

From Hot Atoms to the Stratosphere

Those of us who entered radiochemistry just after World War II were thrown in with scientists only slightly older who had pioneered new techniques and invented new equipment in the Manhattan Project and had also collected lists of potential peacetime experiments. The postwar era had also brought the creation of Brookhaven National Laboratory, with its summer appointments for academics, and in the summer of 1953, I traveled to New Hampshire with the Brookhaven contingent for my first Gordon Conference. I was twenty-five. The authors of the leading radiochemistry text, Gerhart Friedlander and Joe Kennedy, were in their mid-thirties. Even such world-famous "senior" scientists as Glenn Seaborg and Willard Libby were still in their forties. We were collectively young enough that sports dominated the afternoons, and pranks were not unknown. At four in the morning one of the thirties age group led an intrepid band who sneaked into the New Hampton School bell tower to ring the bell in close proximity to Seaborg's room.

Soon after, nuclear chemistry fissioned into radiochemistry, which used radioactivity to study problems involving molecular structure and chemical kinetics, and "nuclear-squared" chemistry, which focused on changes in the nucleus. The Gordon Conferences expanded as well, and most of us began attending conferences that addressed more narrowly defined areas, including radiation chemistry, stable isotope chemistry, geochemistry, chemical kinetics (with radioactive tracers), photochemistry, and isotope effects in chemistry and biology.

In a chance 1970 conversation on an Austrian train after a conference in Europe that was similar to a Gordon Conference, I mentioned to an Atomic Energy Commission (AEC) acquaintance that I had a strong residual Libby-inspired interest in the radioisotopes produced in nature by cosmic radiation. Ultimately, this resulted in an invitation from the AEC to another meeting, one which deliberately mixed two nearly immiscible groups—chemists and meteorologists. There I heard of measurements confirming the presence everywhere in the surface atmosphere of CCl_3F (trichlorofluoromethane), also known as CFC-11. Mario Molina and I explored the fate of CFC-11 in the natural environment, and soon my Gordon Conference of choice became Atmospheric Chemistry, which had gone beyond air-quality and environmental science into the global questions of stratospheric ozone depletion and greenhouse gases. I remained interested in the use of both radioactive and stable isotopes in various milieus, but especially in atmospheric surroundings.

The first Gordon Conference I attended gave me my initial experience in associating with colleagues from everywhere, under conditions where a follow-up discussion to a lecture could happen right afterward in the bar or days later on a tennis court. The GRC parent organization has greatly expanded the opportunities for interactions among scientists, and the organization has been emulated by others around the world. GRC's growth has helped spread this unique approach to sciences other than chemistry, and I am sure it still works as well for newly minted scientists as it did for me fifty years ago.

The first Gordon Conference I attended was the 1983 Condensed Matter Physics Conference (then called Quantum Solids and Fluids). I was encouraged to apply by my thesis adviser, Daniel S. Fisher, who was an invited speaker. I was a graduate student at Cornell University, and I felt an enormous contrast between the atmospheres of GRC and the American Physical Society meeting the previous March, where I had felt quite lost among the thousands of attendees. At GRC everyone ate meals to-gether, socialized, and recreated together (I still vividly remember playing basketball badly in an extremely hot gym), and I felt much more integrated into the community.

I attended this conference again as a postdoc in 1986. Fisher was a conference cochair and invited me to give a talk, which not only was a great opportunity but also gave me exposure that helped in my job search. Like the 1983 conference, the 1986 conference gave an overview of exciting developments in the field.

The Movers and SHAKERS

Management of the Condensed Matter Physics Conference was a bit challenging. Its great breadth provided an avenue to learn about the latest developments in the field; however, the result was that attendees of successive conferences were quite different. The conference chair needed to be proactive and to solicit adequate attendance diligently. The conference also did not have the usual succession of vice chair to chair, which led to communication problems with GRC headquarters. In the early 1990s a sequence of meetings had low attendance numbers. By the time I was elected chair of the 1995 conference, it was on probation and faced cancellation. Vice Chair Sidney Nagel and I spent a lot of time and energy with the GRC staff discussing strategies for developing a compelling scientific program and increasing attendance, and our conference was quite successful. The conference adopted the usual chair–vice chair succession, implemented a biennial schedule, and narrowed its concentration to soft condensed matter physics.

At about the same time the Correlated Electron Systems Conference started up, with its first meeting in 1996. I have attended both conferences, and they each have been quite successful over the last decade. The content of both has evolved and helped define the cutting edge of condensed matter physics.

The Condensed Matter Physics GRC faces issues that are typical of many subfields in the physical sciences, especially in contrast to the biological sciences. The total number of scientists working in the physical sciences is smaller than that of other fields, and many of these scientists work on a range of topics. During a period when my husband worked specifically in a well-defined area of bone research, my own work, for example, spanned various subfields, including complex fluids, condensed matter physics, glasses, granular materials, nonlinear science, quantum computing, strongly correlated electronic systems, and theoretical biology, many of which are covered by Gordon Conferences. For a researcher moving into a new subfield, Gordon Conferences are particularly valuable because they provide an efficient way to learn the latest developments, to meet the "movers and shakers," and to have in-depth discussions. But because the community of attendees is constantly changing, chairs find it difficult to attract the audience, even for exciting scientific programs. This challenge is fitting for a conference of great value, and by facing the issues squarely, Gordon Conferences continue to help define the forefront of the physical sciences in the coming decades.

Susan N. Coppersmith
University of Wisconsin at Madison

Invented

by the Devil

Aart W. Kleyn
Foundation for Fundamental
Research on Matter, Institute
for Plasma Physics Rijnhuizen

Wolfgang Pauli once said that "surfaces are made by the devil"—still an accurate description for the challenges faced by surface scientists. Condensed-matter physicists do not like breaking the symmetry of three-dimensional solids, which was especially true when surface science first came together as a field in the late 1920s following the demonstration of low-energy electron diffraction by Clinton Davisson and Lester Germer and helium atom diffraction from solid surfaces by Otto Stern.

Experimentally, the main difficulty for surface science was the production of clean, well-ordered surfaces, which require using the far-from-trivial technique of making an ultra-high vacuum (UHV). In 1972, when I started working at the Foundation for Fundamental Research on Matter Laboratory of Atomic and Molecular Physics in Amsterdam, UHV required a major effort to maintain. But once the sample was clean and the vacuum held, we could examine the structure of the surface for longer periods.

The Dynamics at Surfaces Gordon Conference, previously Dynamics of Gas-Surface Interactions, is the primary meeting ground for scientists who study the interactions of impinging particles from the gas phase and later the motion of atoms at the surface. In practice, this research involves "contaminating" the surface on purpose, which seemed initially like a counterintuitive experiment for UHV surface scientists. Techniques for studying the dynamics of the gas-surface interaction were derived from experiments on the dynamics of molecular collisions in the gas phase: discussions at Gordon Conferences facilitated this transfer.

By the late 1960s molecular beams had become an important tool to study structure by helium diffraction and interaction dynamics, picking up on studies on the scattering of heavier particles that were halted when Stern left

Germany in the 1930s. Most of the experiments, however, did not yet involve well-characterized surfaces, and many of the experiments gave good insights but not the right data.

A revival came in the late 1970s when UHV molecular-beam surface scattering machines were built in a number of laboratories. At that time I began working as a postdoc at the IBM Almaden Research Center in San Jose, California. In 1980 the first measurements of state-resolved molecular scattering at surfaces were carried out—a novelty for major discussion at the 1981 GRC. For the first time we could observe a molecule rotating and later vibrating when it interacted in a single collision with a solid surface.

The nitric oxide molecule suddenly became the fruit fly of dynamics at surfaces, thanks to its facile state-resolved detection. Nobel laureate John Polanyi triggered a major discussion on "rotational rainbow scattering" at surfaces. I vividly remember him sitting quietly at the edge of the stage while a vigorous debate took place around him. Discussion was interrupted only by a participant asking, "Could the conference chair ask the session chair to stop his debate with the audience and ask the speaker to continue?" Lively discussions of that sort were prevalent in the early Dynamics at Surfaces Gordon Conferences in the 1980s, with the audience often interrupting presentations for long periods.

The field of gas-surface interactions evolved rapidly during the 1980s with several new techniques that for many changed the direction of research interests and laboratory work. The scanning tunneling microscope was invented in the early 1980s, and for the first time surface atoms could be seen and induced to move. For the "beamists" this motion was incredibly slow (one image of the

atom per second was considered fast), since a molecular collision at a surface takes only picoseconds at most. Another major development was the introduction of time-resolved spectroscopy at the pico- and femto-second timescale. Previously, only scattering experiments could detect fast events, but now lasers could track molecules adsorbed at surfaces. Unprecedented views of dynamics at surfaces resulted, and the Gordon Conference, true to its mission of staying on the leading edge of science, changed its name to Dynamics at Surfaces.

Science races on, leaving some unsolved issues behind as attention turns to novel discoveries. Currently, the holy grail of surface science is to make movies of molecular dynamics at surfaces, taken with atomic-scale time and distance resolution. For dynamics at surfaces, new areas of research include more complex surfaces and interfaces, working catalysts, and the flowing liquids of living cells. Many discoveries are only possible through the intense interactions between scientists of different backgrounds. The Dynamics at Surfaces Gordon Conference has been instrumental on this front. My collaborators always enjoy going to a GRC because of the intense interaction with colleagues, a dynamic that is absent at bigger conferences where scientists, especially young ones, get completely lost. Coming together for a week, with a program structured to include large blocks of free time, allows open discussion between scientists. In turn this is a primary driver for the development of open science.

In the late 1970s, during my doctoral studies in chemistry at Ruhr-Universität Bochum, I could not imagine ever attending the Gordon Research Conferences. I had friends in medicine whose professor had applied to attend a GRC oversubscribed by a factor of four and was amazed that he had to wait to hear if he would be accepted. Imagine—a well-known scientist, standing in line, waiting to be selected to attend a conference he had to pay for.

In 1983, when I was working at the German Aerospace Center, the institute director asked a senior scientist if he would recommend attending a new GRC called Laser Diagnostics in Combustion. The New Hampshire college setting did not appeal too much to my superiors, so I volunteered to try it out. I submitted a poster, which was accepted, and I attended my first GRC, the first of many in a variety of fields. A new door was opened. The authors of key articles sitting on my desk became real people within a week. Many of them are still my friends.

Katharina Kohse-Höinghaus
Universität Bielefeld

Since then I have been invited to speak at other Gordon Conferences, and I served as vice chair (1997) and chair (1999) for the Combustion Diagnostics GRC. The 1999 conference led me to put together a book called *Applied Combustion Dynamics*, which was published in 2002. I also submitted a proposal to GRC leadership to refocus the conference in 1997, which was backed by about a hundred personal letters from participants. It was amazing to witness how readily so many scientists from different nations and age groups lent their support to this cause. This outpouring shows how unique and fitted to their respective fields Gordon Conferences are and how critical they are for scientific careers and the advancement of science.

Today I am among the "seniors" who attend the Combustion Diagnostics GRC, and I have followed the paths of more than two generations of talented students and postdocs into academic positions. Some students of my friends from 1983 are professors now whose own students attend Combustion Diagnostics, and some of my former postdocs are among today's speakers, who in turn encourage their students to present their cutting-edge results in poster format. I have also witnessed a steady increase in the number of attendees from around the world and rising numbers of female participants from hard-core science and engineering backgrounds, as evidenced by a comparably shorter line of men for the washroom.

(continued)

WAITING IN *Line*

Speaking of washrooms, I was waiting in line for one at Boston's Logan Airport with other GRC participants when I overheard a well-dressed woman ask another, "Do you see all of those grubby individuals? They camp out in New Hampshire for the summer and are said to be the world's best scientists!" The college environment and a week of sleep deprivation certainly left their marks.

The Laser Combustion Diagnostics GRC was instrumental not only for the careers of those attending but also for the entire evolution of the field—from the first lasers directed into flames to novel sensors in onboard combustion diagnostics. Attending the conference every two years helped fine-tune each participant's achievements. GRC is the only scientific conference series I know of where the discussion periods may be longer than the presentations. The conferences support and create links between people in a way that is both inviting and competitive.

Today laser methods used by combustion chemists and engineers have found their way into other disciplines, including atmospheric chemistry, materials science, and biophysics. I am confident that a new Gordon Conference inspired by these applications will surface from our progress.

> ## Reflections

"My first child (a son) was born one week prior to my first invited GRC talk—just in time for me to attend. His name? Gordon, of course."

Frederick D. Lewis
Northwestern University

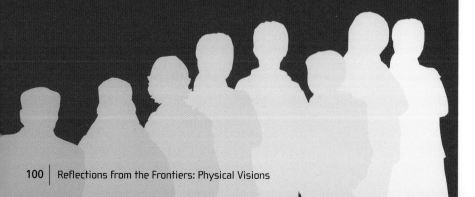

Inspiration
FOR THE WORKING-STIFF SCIENTIST

Where are the frontiers of science? Are they on a sandy bank on the Merrimack River not far from Proctor Academy? Do I hear them in the train whistle of the California coastal mountains? Do I feel them scrambling along hiking trails at Les Diablerets or Mount Sunapee? Do I sense them in a late-night poster session in an elegant Newport mansion or during a late-lunch discussion at Colby-Sawyer College that goes on as long as the topics fascinate? Or do they hide in a civilized glass of sherry at Queens College?

What keeps the frontiers real? What makes them visible, penetrable, and approachable by scientists old and young? Clearly, it is a devotion to science, to its ideas and ideals, its hot topics and burning questions, its leaders and learners, its excitements and frustrations, and its endless challenge and enchantment.

As a working-stiff scientist I find GRC the place to recharge my batteries, to get the juices flowing. This makes some sense: the first GRC I chaired was the Solid State Ionics Conference, which was devoted largely to batteries, and the first GRC I attended was Electron Donor Acceptor (EDA) Interactions, which is all about charge flow.

Charge transfer, energy transfer, and their structural determinants in molecularly organized matter lie at the heart of the science I do, and much of it has been inspired, directed, shaped, and challenged by the EDA Gordon Conference. The openness, the schedule, the size, the informality, the exclusiveness without cliquishness, and the humor make each GRC special and unique. In the timeless words of Walt Whitman, "It avails not, time nor place— distance avails not."

Over the week of the conference new friendships are born and older ones are cemented. New and significant questions shape themselves, and agreements are forged to collaborate on answers to questions formed. I furiously write illegible notes to myself and learn from scientists whose names run from Abruña to Zimmt, hailing from places ranging from Aarhus to Zagreb.

EDA Gordon Conferences are among the oldest, and their shifting focus over the years has illuminated topics from electron transfer in multimetallic systems to light harvesting in photosynthetic structures, from ultraslow corrosion to ultrafast polarization, from molecular and organic electronics to molecular photoionization and photoisomerization, as well as topics like electron charge transfer through peptides, proteins, liquids, solids, and DNA. The GRC has posed, shaped, and focused scientific suggestions: it is critical to the discovery phase of a research program.

Gordon Conferences also shape the community of science. Their origins are in chemistry, physics, and biology, but now GRC topics run seamlessly over the science of objects and processes. The shaping of scientists, however, remains unchanged; as in the past GRC continues to bring together veterans and rookies to play exciting mental games, take part in sport, and foster the personal relationships that are so much a part of the joy of science. I have been at this for three decades, and I look forward to continuing to relish, and to be uniquely rewarded by, Gordon Conferences far into the future.

(continued)

Mark Ratner
Northwestern University

Inspiration

FOR THE WORKING-STIFF SCIENTIST

Change—sometimes radical change—against a background of conceptual continuity and development characterizes the sciences. Gordon Research Conferences are the place where that vibrancy can be felt—the creative tension between what we think we know and what new experiments, models, and understandings hold in the way of future challenge. Because the EDA Gordon Conference meets every other year, my students and I spend about twenty-three-and-a-half months thinking about what we will learn at the next GRC, what questions we will ask, how our understandings will change, and how our appreciation of the fundamental science will be modified.

Science works by self-renewal, which involves realizing the many mistakes we make and reveling in the understandings that the fields achieve. So many of the breakthrough topics, areas, and concepts in my own field would simply not have happened without the GRC evening discussions—punctuated by sugar, ethanol, and occasional ranting—and the impossibly jarring questions that those sessions engender.

Some conferences are far more luxurious than Gordon Conferences, and some are closer to major airports. Most conferences are enormously large and enormously dull in contrast. But Gordon Research Conferences make it more satisfying and more important to do and understand science. They energize, challenge, and reward scientists who attend them, and they crucially aid the ongoing conceptual flow of science in its shaky but steady progress and its intrinsic wonder.

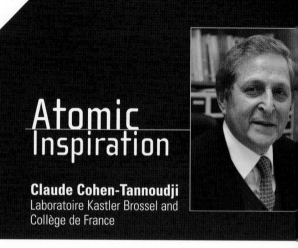

Atomic Inspiration

Claude Cohen-Tannoudji
Laboratoire Kastler Brossel and
Collège de France

Over the last few decades I have attended a number of Gordon Conferences. I also had the honor in 1997 of giving the Alexander Cruickshank Lecture in Atomic Physics.

Gordon Conferences provide much stimulation and gather the best experts of a given field in one place, providing the opportunity for exploration of the depths of their research problems through detailed discussion. The oral presentations are of sufficient length, but enough time is left open during the afternoons for informal discussions and the exchange of ideas. Attention is focused on the scientific problems at hand, without the distraction of having to submit manuscripts for published conference proceedings.

The structure of Gordon Conferences creates a much more fruitful environment than do the large international conferences with thousands of attendees, very short presentations, and many sessions running in parallel. Gordon Conferences have been a source of inspiration for me and have provided the point of origin for many collaborations with colleagues. I have only the warmest thoughts for the seventy-fifth anniversary of the Gordon Conferences, and I hope they will continue with the same success for many years to come.

The MORAL of the STORY

Because the Gordon Research Conferences play a prominent role in defining and discovering the frontiers of scientific research, it is not unusual to hear someone state that GRC has had a significant impact on his or her career. I have certainly said the same about GRC's effect on my own professional path.

I was a postdoc at Northwestern University when I attended my first GRC—the Electron Donor Acceptor Interactions Conference—where I got hints about how to apply for faculty positions. (I also learned all I ever needed to know about long-range electron transfer.) As a junior, untenured assistant professor at Pennsylvania State University (Penn State), I highly valued speaking opportunities at Gordon Conferences. My attendance and presentation at the Electrochemistry GRC was important for establishing connections with people interested in the niche I was carving out at the juncture of nanoparticles, interface science, and biosensors. During my tenure as a professor of chemistry at Penn State I became a frequent attendee at the Chemistry at Interfaces, Clusters, Nanocrystals, and Nanostructures and the Bioanalytical Sensors conferences.

GRC also played an important role in my choice to forsake academia and tenure at Penn State for the thrills (and chills) of entrepreneurship. A talk I gave at the Bioanalytical Sensors GRC struck a chord with an industrial attendee from Affymax, Inc., and led to a collaborative research program with my lab, which, in turn, opened the door to my consultancy with the company. As interest in our work at Penn State grew among biotech and pharmaceutical companies, I presented a business plan to the now-former CEO of Affymax , Gordon Ringold, that laid out possible commercialization opportunities for the new technologies we were developing. He liked it, but pointed out that my dual role as Affymax chief science officer and full professor at Penn State was a nonstarter to prospective investors. I had to pick one or the other. So I chose to stay with Affymax, and the rest, for better or worse, is history.

The moral of the story is that the Gordon Conferences provide a unique forum to present and hear about science at the absolute cutting edge. For me as both a speaker and a listener, the GRC experience has prompted important career consequences spanning academia and industry. Whether for technical education or career development, I cannot think of a better way for science professionals to spend time out of the lab or office than attending a Gordon Conference.

Michael J. Natan
Nanoplex Technologies

Offering
Unguarded
Speculation

Harry Atwater
California Institute of Technology

The first Gordon Conference I attended was called Physical Metallurgy; I was a postdoc at Harvard University. I was definitely thinking about my professional life as a scientist and teacher, since I was still young and unestablished. This was the first time I had ever been to a conference that was not a "hotel conference"—the kind of conference where people dressed up in suits and ties. It was revealing to me that you could actually approach senior, very visible people informally, go sailing with them, and talk to them at the same time. People were very approachable and willing to offer unguarded speculation.

When I have subsequently organized conferences, I think back to that very first Gordon Conference: it set the gold standard for what a scientific meeting should be in terms of the quality of the program, the quality of the discussion, and the way that young people can interact with senior scientists.

I was a little suspicious when I first went to a Gordon Conference that the time in the afternoons would be wasted. But I found it to be quite the opposite; the information density of the whole meeting was much higher than that of other scientific conferences.

Interacting with the senior people in metals at the Physical Metallurgy GRC was especially important for me because I got into materials science through the back door from physics and electrical engineering. Thanks to GRC, I was quickly exposed to new levels of analytical thinking about kinetic processes and materials. It was amazing to see a community of people who, without even writing down any equations, could make detailed models in their minds of things that electrical engineers had treated very descriptively. There was an implicit sense that this was a special meeting for the field. For me, attending GRC instilled a sense of rigor about modeling transformations in materials that had a significant impact on my research career.

The other great thing about Gordon Conferences is that there is a common vernacular, so the conversations can go very fast. People do not offer a lot of preparatory and didactic remarks, nor are they needed for scientific discussion to move ahead. You can start a conversation, safely assuming that everybody knows the events of the past year. Physical metallurgy is a field small enough and the focus of the conference is sharp enough that the discussion can move quickly. Beyond GRC, I have only had the same kind of energetic experience when working closely with international collaborators on a specific project. GRC introduced me to a format where everyone is already up to speed on an issue; you jump right in, and you immediately go at it at a very high level.

The
New
Generation

A Venue for Integrative Scholarship

Arthur B. Ellis
National Science Foundation

Like other materials scientists I greatly admire the elasticity of the Gordon Research Conferences. As a faculty member conducting multidisciplinary research in the chemistry department of the University of Wisconsin at Madison, the Gordon Conferences enabled me to share ideas with chemists and scientists from other disciplines. Attendance at the Solid State Chemistry, Inorganic Chemistry, and Electrochemistry conferences helped my research program evolve over nearly thirty years in academia. Moreover, the network of professional contacts and mentors I developed through Gordon Conferences has been a valued source of advice and inspiration for my entire professional career. The Gordon Conferences exposed me to international colleagues whose perspectives have broadened my outlook on the scientific enterprise and to young colleagues who have affirmed my confidence in a bright future for science. From conversations with many colleagues across a multitude of scientific and engineering disciplines, I know that my perceptions are widely shared.

GRC supported my interest in integrating results from the chemistry research enterprise into the undergraduate curriculum. From 1977, when I began my career as a faculty member, to the early 1990s, neither the content nor the pedagogical methods of introductory college chemistry courses changed much. In 1990 I started working with colleagues across the country to integrate solid-state chemistry into these courses. With support from the Camille and Henry Dreyfus Foundation and the National Science Foundation (NSF) we created a suite of instructional materials to accomplish this objective. The Solid State Chemistry and Inorganic Chemistry conferences proved superb venues for introducing these instructional

materials to other colleagues. GRC chairs graciously made time and space available during otherwise unscheduled afternoon sessions so that we could share the materials with our colleagues. We thereby obtained expert feedback that ultimately led to better products and enhanced the visibility of our effort.

In 1992 I had the good fortune to attend the first Science Education Conference. The participation of many leading scientists signified that the research enterprise was awakening to the need to update our science courses. This meeting was both invigorating and transformational: it brought together disparate communities and demonstrated that by continuously infusing science curricula with the fruits of research and technology, the educational enterprise would have the same vitality as the research enterprise. It was also fortunate that Angy Stacy of the University of California at Berkeley attended this meeting. The two of us concluded that we should try to organize a GRC like the Science Education Conference, but one that focused specifically on chemical education. The GRC board approved the proposal we submitted, and the Innovations in the Teaching of College Chemistry Conference was launched in January 1994. The backing of GRC and the involvement of many leading chemistry researchers gave the revitalization of the chemistry curriculum a credibility that would have been difficult to achieve otherwise. Furthermore, such backing enabled us to attract funding from the Dreyfus Foundation, the Research Corporation, and the NSF, which helped ensure the involvement of broad segments of the higher-education chemistry community, including two- and four-year institutions, as well as publishers and producers of instructional materials.

Laudable Sins

This yeasty mixture of stakeholders in undergraduate education led to an exciting inaugural GRC.

Since joining NSF in 2002, I have worked with colleagues in the Division of Chemistry and across the foundation to advance the frontiers of discovery, integrate research and education, and broaden participation. The objectives of GRC make it a natural intellectual partner for the NSF: GRC provides a forum for cutting-edge research and education that engages an inclusive spectrum of participants spanning a variety of disciplines, institutions, demographic groups, professional development stages, and state and national borders. Our nation's scientific enterprise has been greatly enriched by this synergy, and I wish the GRC continued success in its important work.

The views expressed are those of the author and do not necessarily reflect the views of the National Science Foundation.

George Pimentel once made a statement that became a classic: "Teaching is to research as sin is to confession; you can't have one without the other." That idea has always been part of the GRC tradition: researchers in a discipline educate each other at Gordon Conferences.

During my involvement in GRC, science education was recognized as critical to the development of future generations of researchers. Education was also part of the heritage of GRC as instituted by its founder—the first editor of the *Journal of Chemical Education,* Neil Gordon. Alex Cruickshank clearly understood this. He once let Dick Holm and me, when we were graduate students at the Massachusetts Institute of Technology, sit in on a morning session of the Inorganic Chemistry GRC at the Holderness School in Holderness, New Hampshire, in 1958. Our adviser, Al Cotton, was an invited participant, and Bob Parry and Fred Basolo were in attendance. For the first time Dick and I were able to associate faces with the names of leaders who were creating a renaissance in inorganic chemistry. Basolo and Parry encouraged chairs of the Inorganic Chemistry Conference to urge speakers to describe the impact of their work on the development of inorganic chemistry.

John P. Fackler, Jr.
Texas A&M University

I took my place as the Inorganic Chemistry Conference chair in 1979 and was subsequently elected to the GRC council. During the next nine years I gained a great appreciation for the legacy of Neil Gordon. Alex Cruickshank became more to me than just a competent director: he also became a friend and confidant. It was a great personal delight that the Alexander M. Cruickshank Lectures were created while I was on the board of trustees. However, the board also had to discuss the consequences and processes associated with Alex's eventual retirement.

The board debated two important matters that have influenced GRC: the role education conferences might play in the structure of GRC and the value of holding conferences outside the United States, particularly in Europe. Bob Grasselli helped the board decide on conference sites in Italy and Germany. When I was chair of the board, we encouraged a conference on science education, with Alan Cowley and the late Paul Saltman taking leadership roles. A GRC on chemical education called Innovations in College

(continued)

Laudable Sins

Chemistry Teaching was held in 1994. The same year Gerhard Pohl and I organized the first education conference in Europe called New Visualization Technologies for Science Education, held in Irsee, Germany. The conference was truly international, with scientists and science educators attending from all parts of the world. I have many fond memories of this conference, set not far from Munich and scheduled in late September just before Oktoberfest.

Although I have most often attended the Inorganic Chemistry Conference, I have also participated in the Metals in Biology, Electrochemistry, and Organometallic Chemistry conferences. Involvement with four science education conferences, however, has convinced me that Pimentel's statement requires encouragement of a positive, supportive relationship between those whose primary focus is on discovery in science and those who educate society about its worth. The Visualization in Science Education Conference has brought Nobel laureates to the same platform as science museum directors and science writers. At this conference I have also learned that many countries around the world have successfully created environments where discovery and education in science are respected by society. Interactions between science teachers and researchers are indispensable links for a strong, educated society in the modern technological world, and I am hopeful that GRC leadership will continue to build on these relationships.

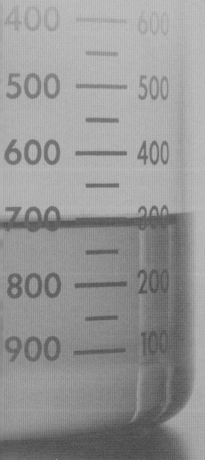

Contacts, Collaborations, and the *Buddha*

I was a graduate student at the Weizmann Institute of Science in Israel when I first attended the Chemistry and Physics of Isotopes Gordon Research Conference. I was then doing research on secondary isotope effects. To me, established scientific names like Jake Bigeleisen and Jack Shiner carried the same weight and authority as equations and theories. Spending five days with people whose articles and research I revered was an unbelievable experience. I had never imagined it possible to talk with the leaders in my field. My first GRC experience and the contacts I made shaped my future career.

The structure of the Gordon Conferences allowed me to spend free time in the afternoons having informal discussions while climbing mountains or sailing the lakes of New Hampshire with Jake Bigeleisen at the helm of the boat; we were able to continue discussions on the size of isotope effects, tunneling effects, or gas-phase isotope effects. (Jake was the Buddha of isotope effects.) Conversations lasted late into the night, typically until 2:00 or 3:00 a.m., in a room set aside for this purpose. These experiences were more valuable than I as a young researcher could have possibly imagined.

The Gordon Conference format, which is surely the best model for getting the most out of a conference, helped me develop valuable relationships. When I returned to Israel, I felt I could call upon these colleagues — friends I had made at the conference—for input. Without the Gordon Research Conferences I could never have developed the research I was working on.

When I finished my Ph.D., I needed to look for a postdoc position. Although a foreigner, I had a distinct advantage over other applicants because of my participation in the Isotopes Gordon Conference. I applied directly to Franklin A. Long, whom I had met at the first Gordon Conference I attended. I told him I was interested in working with him on isotope effects and proton transfer reactions. Without this valuable contact I do not know where I would have ended up.

I rarely missed an Isotopes Gordon Conference because I saw them as the greatest resource I had for pursuing my research. Even after I decided to shift my career and became the head of the Institute for Science Education and Science Communication, I continued to keep abreast of the field of isotope effects by attending the conferences.

After my career transition I was invited to talk at the first New Visualization Technologies in Science Education Gordon Conference in Irsee, Germany, in 1994. I have found these conferences to be as enlightening and beneficial as the Isotope Conferences. I participated in all the Visualization Conferences that followed (now called Visualization in Science and Education). I formed new relationships, and my group of five was awarded a mini-grant to study visualization in the United States, Russia, Kenya, Eritrea, and Canada. This year five of us who participated in the Visualization GRC are studying the effects of culture on science visualization. I seriously doubt this collaboration would have occurred had it not been for GRC.

Zafra Lerman
Institute for Science Education and Science Communication at Columbia College, Chicago

importance, that the Gordon Conferences had apparently never held any meetings on science policy. "Well, propose one," he said and told me how. I followed his suggestion: the Gordon board approved the proposal and a meeting on the topic was held in 2000. Two more followed, in 2002 and 2004, with a fourth scheduled in 2006.

I subsequently discovered that GRC had in fact sponsored a meeting in 1968, largely devoted to science policy, called Science, Technology, and Economic Growth, but that one breathed only for that year, never to inhale again. Policy as a Gordon topic returned to its cove in yonder hills, except for lectures or allusions to it in various other Gordon Conferences over the years.

A quick historical review shows that the seeds of today's astonishing era of discovery were sown by World War II, when the United States' scientists and engineers were called into service to help win it. The nation's researchers turned out a spectacular stream of innovations that included synthetic rubber, new drugs, surgical techniques, electronic espionage systems, early computers, radar, and most ominous of all, atomic weaponry. After the war Vannevar Bush, the MIT engineer who managed the wartime research effort, produced the revolutionary report "Science—the Endless Frontier," which set the framework for government support of science for both peaceful and military purposes. The ultimate result was massive financial support for research in all fields of basic science and, in time, for engineering and the social and behavioral sciences as well. At the same time scientists, because of their work on the atomic bomb, were forced to think about and engage themselves in the social, ethical, and moral consequences of their efforts.

expanded enormously, research universities became active lobbyists, and every major scientific and engineering society found itself compelled to establish policy units. As early as the mid-1960s public policy concerning science and technology was becoming an academic discipline in its own right.

Many issues are ripe for policy research, acute historical analysis, and incisive topicality, including industrial influence on academic research behavior; global change and its social and political consequences; effective government structures for good policy making; stem-cell research and its ethical considerations; energy policy in a time of dwindling oil availability; the vague but critical field of "homeland security" in the United States accompanied by a controversial research agenda and a host of restrictions on information and immigration; the ethical and social responsibilities of industrial research; the comparatively low level of science literacy in the United States; and the profusion of controversies in medical and agricultural genetics. The list goes on and on.

So much for background. To organize the initial meeting, I needed partners, and so I called on Marty Apple, then president of the Council of Scientific Society Presidents, and Daryl Chubin, then on the staff of the National Science Board. Marty volunteered as cochair and Daryl elected to be vice chair (which meant he would chair the conference's second meeting). Together we mapped out a program that by its diversity would determine what would work within the GRC format and what would not.

The Science and Technology Policy Gordon Conference in 2000 began with a keynote session on future issues across the spectrum of science and technology policy. Following were sessions on the implications of information science, science education, science communication, international relations in science and technology, the ethical issues in molecular genetics, trends in defense research, and the basis of protest movements against the Western domination of science and technology. The topics stretched across many issues and disciplines and were seen by many as too light on the research side and too wide in subject matter. Still, much was learned, and the conference was under way.

The second conference was held in 2002, organized by Daryl and his cochair John Elter of Plug Power Corporation, and by vice chair Jane Maienschein, a historian and philosopher of science at Arizona State University. Jane joined with Susan Fitzpatrick, vice president of the James S. McDonnell Foundation, to cochair the 2004 conference. The 2002 meeting, while more focused and interactive, was disappointingly small. Clearly, Jane and Susan had their work cut out for 2004. By dint of hard work and outstanding fundraising, the conference, held at Big Sky Resort in Montana, was a smash. Rich with students, bulging with posters, and oversubscribed, the conference was diverse in makeup and focused subject matter.

The 2006 conference is well along in the planning stages and will be cochaired by biologist and bioethicist Fred Grinnell of the University of Texas Southwestern Medical Center at Dallas, and Skip Stiles, an independent policy consultant and former staff member of the late U.S. Representative George Brown of California, and co–vice chaired by political scientist David Guston of Arizona State University and bioethicist Rachel Ankeny of the University of Sydney.

Conference organizers are aiming to increase international participation, which was modest up to 2004. GRC leadership also began rethinking the criteria for properly evaluating Gordon Conferences of the policy type, which should include issues of idea generation, the pertinence of case studies, the reporting of scholarship, and the shaping of research agendas.

One critical hope is that more of the Gordon Conference community will participate in the evolution of policy-type conferences. It is the work of these scientists that has set the character and tone of the modern age, and it will be their thoughts and their discoveries that will set the agenda for the policy community. Science and Technology Policy participants believe that while the first goal of the conference is to promote excellence in the policy community, the discussion is ripe for the presence of the entire community of Gordon Conference participants.

Raising the
NEXT GENERATION

Starting a new career is similar to joining a game late without knowing the rules, when players already know each other's strategies and beliefs. They are wary of outsiders but understand that the only way of adding zest is to bring in new competitors.

Graduate students' sources of contacts often are their thesis advisers' colleagues and collaborators. In my later years as a graduate student I realized I would not have the luxury of my adviser's introductions forever and that I needed to create my own network. But the question was how to go about doing it.

Juana I. Rudati
State University of
New York at Stony Brook

It is not easy for young scientists to get to know more advanced scientists who might have a say in who gets important grants or which results get published. Attending major meetings exposes young scientists to hundreds, maybe thousands, of more experienced scientists, but only quick conversations can be achieved there. Most attendees are busy preparing talks, interviewing future colleagues, or discussing results. Opportunities for undisturbed chats are few and far between. Meals are natural opportunities for relaxed conversation; large meetings, however, cannot possibly accommodate everyone with meals, so attendees go their separate ways—with their "usual crowds."

This is where Gordon Conferences make a big difference. Because the number of attendees is limited, it is possible to get to know everyone. Meals are included as part of the conference and are scheduled in a cafeteria filled with large round tables. It forces conferees to reach beyond their comfort zones and explore new relationships. In turn, those relationships can provide new links in a scientist's network as well as future research collaborators. There are few people I work with today that I did not meet at a Gordon Conference.

(continued)

Gordon Conferences are scheduled to provide plenty of time for stress-free exploratory conversation and unrestricted activities. The conferences I have attended have provided time to go canoeing or to ponder research questions with fellow graduate students, scientists, or professors who are tired of working on grant applications. During those unrestrained discussions no research idea is too crazy. Free time is also an opportunity for career advancement. Going for an organized hike with a prospective postdoc adviser can act as a sneak preview of the dynamics of such a relationship. People looking for postdocs and junior faculty can get more thorough impressions of candidates. Activities also bring renewed energy necessary for the evening talks.

Gordon Conferences run only one session at a time, which has allowed me to talk comfortably with many authors of papers I have read. There was no need to cut conversation short to go to another session or to sacrifice attending a talk in order to continue conversation.

Most important, Gordon Conferences help to build camaraderie. The same crowd shows up at every meeting, with a few new people each time to add spice to the conversation by bringing new perspectives to ongoing discussions and new research results for analysis. Furthermore, GRC provides an opportunity to talk not only to professors but also to their students who work on experiments day in and day out.

I have felt welcome since the first Gordon Conference I attended. I applaud the organization for maintaining the perfect combination of conference features that make the experience a success for both long-time attendees and new generations.

> Reflections

" The nature of the genetic code was an important topic at the 1961 Nucleic Acids Conference. Joe Sambrook (from Severo Ochoa's laboratory) gave a paper on some experiments identifying codons. When asked about how the experiments were done, he said that he was not at liberty to say. There were boos and catcalls from the audience. That evening Sambrook could be seen in the only phone booth (presumably calling New York). The next day he asked for extra time to speak, explaining that he now had permission to discuss the methodology. "

Bernard Strauss
University of Chicago

Transcendental GRC

Principles

A Time of *Change*

Alan H. Cowley
University of Texas at Austin

I was introduced to the Gordon Research Conferences in the mid-1960s when I attended an Inorganic Chemistry GRC, a career-influencing event in every respect. I had never attended a meeting with such lively, abundant discussion. Moreover, the informal ambience encouraged both new friendships and meaningful interactions with several distinguished inorganic chemists whose work I had read and admired. I became an aficionado of GRC, and with Alex Cruickshank's encouragement, I got involved with the organization's management. I was elected to the GRC council for 1984 through 1987 and to the Selection and Scheduling Committee for 1988 to 1989.

Serving on the GRC board of trustees from 1989 until 1998 was a particular privilege and pleasure for me. The organization underwent significant change during this period. I served on the board for nine years—a record I share with Jim Florini. In fact, my tenure comprised two consecutive terms—an anomaly that arose because the limit on board terms changed from three to six years during my first term. This change was prompted by the realization that because GRC had grown significantly in size and complexity, three years was not enough time to "learn the ropes." An elaborate changeover system was devised, largely by Tony Trozzolo, to account for board terms that had started in different years, and Jim Florini and I had the option to run for subsequent six-year terms. Alex Cruickshank's retirement and Carl Storm's subsequent appointment as the new director in 1993 were other significant changes during this time. I am indebted to both of these gentlemen for their abundant advice and warm friendship.

A major theme that evolved during my tenure as a trustee was internationalization. Even though conferees from outside the United States constituted more than 20 percent, in the early 1980s the board debated whether the organization regarded itself as a U.S.-only institution or whether it was more international in scope and philosophy. Beginning in the late 1980s, under the leadership of Bob Grasselli and John Fackler, the latter view prevailed, with its implementation being a concern of the board of trustees in the early 1990s. Efforts to internationalize took a number of forms, including visiting many prospective overseas conference sites with Carl Storm and dealing with various diplomatic issues. I have particularly vivid memories of a conference in Irsee, Germany, at which the board met with distinguished scientific representatives from Sweden and the then-member states of the European Union. Our main objective was to avoid any misunderstandings and to set forth reasons for holding Gordon Conferences in Europe. Although the meeting got off to a wobbly start, it finished amicably, and a cooperative arrangement was forged with the then-fledgling European Research Conferences.

During my first board term many had concerns about the diminishing number of potential scientists in the United States. The board agreed that GRC could and should play a role in helping to reverse this trend. In this context, and mindful of Neil Gordon's contributions to science education (as the founding editor of the *Journal of Chemical Education,* for example), the idea of holding a Science Education Gordon Conference emerged. My

Pursuing a *Passion* & *Lost Keys*

As a graduate student in 1970 I was excited to attend my first GRC, Chemistry and Physics of Solids, at the Holderness School in Holderness, New Hampshire. The conference was attended by many accomplished scientists whose names I knew from papers and group discussions. I must admit I felt rather ignorant, but everyone made me feel welcome. I even got to play in the poker game with the stars (I lost). The amount of time permitted to discuss what people were doing as of yesterday, with no distractions save some afternoon swimming, was incredible.

I have been attending GRCs ever since—having been to more than thirty by now. They are the only meetings where there is time to form and discuss ideas—some crazy, some not so crazy. I love the mix of students, young faculty, and old-timers like myself. A meeting highlight is always the evening poster sessions, which are small enough to discuss and take in all the work—mostly unpublished and sometimes still "raw." It is hard to find other conferences that have the same feeling of dynamism—probing and advancing the frontiers of one's own field or related fields. I always get home feeling rather exhausted from so many late nights and a little too much wine.

Frank DiSalvo
Cornell University

One summer I was invited to speak at three Gordon Conferences in New Hampshire. Including a week of vacation, my family spend a whole month at Squam Lake. What a summer! I cannot imagine finding it worthwhile to attend three conferences in a month if they were not Gordon Conferences, summer or otherwise. The ideas at the Gordon Conferences were just popping: it was very exciting. I was working at AT&T Bell Laboratories then, and, even though Bell was a terrific environment for building collaborations, I wanted to stay longer in New Hampshire to keep learning from and talking about results and ideas with such an accomplished and diverse set of scientists.

I have many memories of the fun and the little misadventures we had as well. One summer I drove a group to a lake about fifteen miles from the conference site for an afternoon of swimming and talking. I put the car keys in the pocket of my swimming trunks, and—you guessed it—they fell out while swimming and were lost. We all eventually got back to the conference, but I had to wait two days to get a new key for the car. On another occasion I took a group hiking in the mountains and got the group lost. We all got back after dinner rather tired and hungry. The staff took pity on us and brought us leftovers to eat, so it all worked out in the end. Somehow recalling these and many other adventures still tickles me. These events as well as the science have helped me make many friends.

colleagues thought I should organize the first such conference. Being a hard-core research type, I was somewhat daunted by this prospect but fortunately was inspired by a *Newsweek* article featuring the late Paul Saltman, a distinguished educator and teacher. Together Saltman and I assembled a lively mixture of distinguished speakers and discussion leaders who represented the research and teaching communities. Much vigorous discussion took place, with the total attendance exceeding 160. Today science education remains an integral part of the GRC portfolio.

> Reflections

" By choosing different areas to emphasize each year, the Condensed Matter Physics Gordon Conference helped counterbalance the centrifugal forces endemic to rapidly growing fields in the process of splitting into subfields. Much cross-fertilization resulted. "

Morrel H. Cohen
Rutgers University

(continued)

Pursuing a *Passion* & *Lost Keys*

Even now, I cannot recall other times in my career when the number of ideas discussed or inspired were so many in such a short period. I always had plenty of new things to try in the lab when I returned home from Gordon Conferences.

I am not particularly fond of committees, but because I profited so much from the conferences themselves, I happily agreed to serve on GRC's Selection and Scheduling Committee. There I met scientists from disciplines I knew nothing about. They were interesting people and great fun to work with. In 1994 I was fortunate enough to be elected to the board of trustees, and I was privileged to serve as chair in 1997. The trustees are indeed a wonderful group—all talented scientists who feel that Gordon Conferences were very important to their careers and fields. The combined experience and wisdom of the group is impressive. Of course, our meetings were always held in interesting places, so we all enjoyed ourselves immensely.

If Gordon Conferences did not exist, we would have to invent them. They are meetings for scientists, run by scientists, doing what they like best—pursuing their passions with similarly minded scientists. This philosophy has clearly endured. Gordon Conferences are now held around the world and are sometimes imitated by other organizations. Because of volunteer scientists who listen and have assembled a superb staff at the helm, GRC is dynamic and continually evolves to meet the needs of the scientific communities it serves. On the occasion of the seventy-fifth anniversary of GRC's founding, I wish us all continued success. Together we will always have a bright future.

> ## Reflections

" I remember watching some wonderful tennis matches between famous scientists. Also I remember that there were only about two public telephones and people would be lined up to telephone their labs at home about hot new leads and ideas."

Dorothy Bainton
University of California
at San Francisco

TOUR DE FORCE

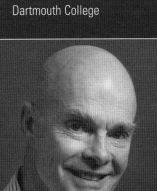

David M. Lemal
Dartmouth College

One of my most memorable experiences during my long association with the Gordon Research Conferences was in 1978 when I was chair of the board of trustees. Alex Cruickshank, who I had known for many years, was a warm, friendly gentleman, and I admired his able and personable leadership of the conferences. Alex arranged for conference meeting sites with a handshake rather than with paper contracts.

Alex telephoned me one day in January, reporting that a headmaster had called to report that his school was bailing out of its commitment, made several months earlier, to host a full schedule of conferences in the upcoming summer because it was too costly. Faced with this egregious breach of a firm agreement very late in the conference planning process, Alex swung into action. Within forty-eight hours of the headmaster's phone call, Alex had arranged for alternative homes for all the conferences. Then he arranged a meeting for us with the headmaster.

On the appointed day at 9:00 a.m. we sat down with the business manager and the headmaster of the school, who began by complaining bitterly about what a losing proposition the conferences were for their school. Alex, without ever raising his voice or seeming piqued, began asking questions about the school's finances. The headmaster deferred to his business manager for answers, whose replies were usually in effect, "I don't remember, but I can look it up." After each of these responses Alex then proceeded to answer his own question, and in the process presented an astonishing knowledge of the details of the school's financial situation and dealings.

I do not know how he acquired all the information, but his performance was truly a tour de force. He made a convincing case that, far from suffering, the school enjoyed great financial benefit from the opportunity to host the Gordon Research Conferences. Close to noon, the headmaster penitently asked Alex whether he would be willing to have the school host the conferences after all. It was agreed, and the school remains a conference site today.

Alex had revealed a side of himself I did not know existed. That gentle exterior hid a tough, canny Yankee who was remarkably knowledgeable and resourceful. My respect for him took a quantum leap, and I came to understand more fully why the Gordon Research Conferences had become the premier international meetings for chemistry, biology, and physics.

LASTING Memories

Kathleen C. Taylor
General Motors, retired

My first GRC experience was the Catalysis Gordon Conference I attended as a young scientist in the early 1970s. I recall the many great names in the field that I had the opportunity to meet, the lively scientific debate, and of course the beauty of New Hampshire in the early summer. My lasting memories of those years are of people widely recognized for their scientific achievements and industrial applications and of friendships created that endured throughout my career.

I was excited to attend, especially because acceptance to the conference was very competitive. Usually only two people from a given company would be selected to attend; the companies knew this and would only submit two names. I had joined the General Motors Research Laboratories in 1970, where I was involved in the early catalysis research related to the automotive catalytic converter. (Prior to that I had done my graduate work under Robert Burwell, Jr., at Northwestern University and my postdoctoral studies with Charles Kemball at the University of Edinburgh.) *Catalyst* is now a household word, as the catalytic converter has since brought about remarkable reductions in automobile emissions. The development of the automotive catalytic converter exemplifies how the field of catalysis has continued to reinvent itself in response to new needs, opportunities, and discoveries, and the Catalysis Gordon Conference is where it happens.

I recall from early years of the Catalysis Conference a long list of outstanding scientists: my most vivid memories are of Robert Burwell, Michel Boudart, Paul Emmett, Herman Pines, Vladimir "Val" Haensel, R. Paul Eischens, and W. Keith Hall, who was famous for sustaining late-night scientific discussions. The Gordon Conferences had their summer headquarters at Colby-Sawyer College in New London, New Hampshire. The Catalysis Conference traditionally covered major and emerging research themes instead of focusing on a single area of the field. Catalytic science and applications were addressed in the search for new understandings. Speakers were drawn from the broad community of scientists and engineers from academia, industry, government, and technological societies. Additional conference activities included swimming, sailing, walks in the White Mountains, tennis, and soccer matches between scientists from the United States and those from elsewhere in the world. Other long-standing and memorable traditions were the Thursday-evening lobster banquet, the song fest, midnight swims, and pizza feasts.

I chaired the Catalysis Gordon Conference in 1993. I had already served on the Selection and Scheduling Committee for several years and served as a board member from 1991 to 1996 and board chair in 1994. In these years the number of conferences in the United States was rapidly expanding, and so the board decided to start a series of conferences in Europe as well.

rewarding
Moments

I still have strong memories of my first Gordon Conference in 1963 at Kimball Union Academy. It was wonderful to talk one-on-one with leading scientists in my field. At the time I had no idea that the Gordon Conferences would play such a large role in my life and provide some of my most rewarding moments. I am particularly pleased at the continuing success of the Hormone Action GRC, which I started in 1969; I am nearly as proud of it as I am of my two daughters. I also facilitated the foundation of the Peptide Growth Factors and the Myogenesis Gordon Conferences.

James R. Florini
Syracuse University

One of the best features of Gordon Conferences is access to its great speakers. At a Peptide Growth Factor Conference in the mid-1980s, Nobel laureate Stanley Cohen (a towering fixture in the growth factor field, having been the first reporter of nearly every major growth factor discovery) gave the keynote Sunday-night lecture. At breakfast Monday morning I happened to be sitting at the table with Stan and several other people when a young woman sat down with us. After some hesitation she joined in the conversation, asking Stan a few questions about his talk. As is his very gracious habit, he answered her fully and frankly. A bit later, when we were leaving the dining room, she turned aside to me and said, "I never dreamed that I could ask Stanley Cohen why he did an experiment!" It turned out that she was a brand-new postdoc attending her first scientific meeting. To me, that is the essence of a Gordon Conference—a new postdoc asking a Nobel laureate about his experimental choices.

(continued)

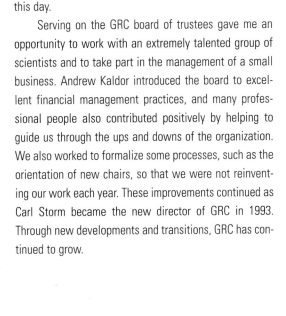

I first met Alex Cruickshank, the man who put it all together, in the early years of the Catalysis Conference at Colby-Sawyer. Alex was always forward-looking. The trustees made the decisions, but Alex always gave them good choices. He let the science happen on its own and recognized the needs of scientists to meet and share their work in the Gordon tradition. Alex's dedication to the conferences was phenomenal, and his skill in working with the host schools was legendary. His ability to embrace the new, yet keep the unique core character of the Gordon Conferences intact, was very important to his success. My time serving on the board of trustees included the years just before Alex retired, and Alex remains a dear friend to this day.

Serving on the GRC board of trustees gave me an opportunity to work with an extremely talented group of scientists and to take part in the management of a small business. Andrew Kaldor introduced the board to excellent financial management practices, and many professional people also contributed positively by helping to guide us through the ups and downs of the organization. We also worked to formalize some processes, such as the orientation of new chairs, so that we were not reinventing our work each year. These improvements continued as Carl Storm became the new director of GRC in 1993. Through new developments and transitions, GRC has continued to grow.

Rewarding Moments

Just as important, Gordon Conferences make possible the transfer of critical scientific advice through informal communication. In the summer of 1980 *Time* magazine sent a reporter to New London, New Hampshire, to do a feature. For whatever reason Alex Cruickshank sent her to nearby Andover, where I was chairing the Biology of Aging GRC, and I happened to have a story that illustrated the importance of informal communication at Gordon Conferences. At the same conference two years earlier Ed Masero of the University of Texas, San Antonio, asked me during a morning coffee break what I was planning to do next. I outlined projected experiments on the effects of age on amino acid transport into isolated diaphragm muscle (which is one way to measure protein anabolic responses). Ed, an experienced physiologist, pointed out that this approach, while fine for the thin muscle in young rats, would not be appropriate for the much thicker muscles of the twenty-four- to twenty-eight-month-old animals used in studies of the effects of aging. In these muscles the interior of the diaphragm becomes anoxic and releases many degradation products, completely invalidating any resulting measurements. As I told the *Time* reporter, who quoted me in the article, that information probably saved us a year of wasted work, and the situation illustrated what a Gordon Conference can make possible.

Gordon Conferences also reap incredible personal rewards. In 1981 I was elected to the Selection and Scheduling (S&S) Committee, although at the time I knew little about its duties or even its existence. For various reasons I attended every fall S&S meeting from 1981 to 1999, as well as nine board of trustees meetings. At a wonderful surprise ceremony at the 1999 S&S meeting I said, "When I went to my first S&S meeting in 1981, I had very little idea of what I was to do, and I certainly didn't think I would be doing it for the rest of the century." At that event I was given a beautiful Tiffany crystal bowl inscribed "Professor James R. Florini; 18 Years at the Frontiers of Science—Gordon Research Conferences." Those eighteen years included some of the most enjoyable interactions I have ever had. It was wonderful to work with such able, accomplished, distinguished, and delightful people on what we all considered the best scientific meetings in the world.

The Gordon Research Conferences have affected my professional life in a very personal way. In 1963, near the end of my first year on the faculty at the University of Utah, I was involved in preparing my first research proposal on the effects of high pressure on the properties of materials. Professor Baker, chair of the mechanical engineering department, on receiving an announcement of the upcoming Gordon Conferences, asked if I would like to attend the Research at High Pressures Conference that summer.

Attending the conference, held at Kimball Union Academy in Meriden, New Hampshire, remains one of the highlights of my career. Also in attendance were Harry G. Drickamer, George B. Benedek, and other researchers whose works I had came across in my literature search. I was starstruck and used what I learned at the conference to revamp my proposal before submitting it to the National Science Foundation (NSF). The resulting award was the first of nearly forty years of uninterrupted research support from NSF; the research it spawned was fun, interesting, and rewarding. Three years after receiving the first grant, colleague Wayne S. Brown and I successfully submitted a proposal to the Department of Defense to explore the properties of rocks under high pressures using equipment I had developed. Not suitable for a university environment, this applied research was put to use by a company called Terratek founded in 1969 for geotechnical testing for the gas and oil industries.

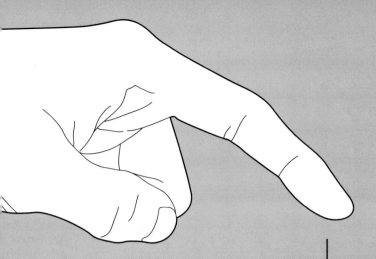

Professional *Meets* PERSONAL

K. Larry Devries
University of Utah

In 1969 I was looking for new research directions, and I had ideas (based on pioneering work by M. L. Williams) on how fracture mechanics and polymer science might be used in the analysis and design of adhesive joints. I applied to attend the Science of Adhesion GRC. Dominant leaders in adhesion and adhesive science, such as Robert L. Patrick, Louis H. Sharpe, Carl A. Dahlquist, and Alan N.Gent, were participants. Once more a Gordon Conference propelled my research group into an exciting area. I attended most of the next ten or so Science of Adhesion conferences. I was honored to be elected chair for the 1975 conference, where a group got together and, under the leadership of Gent, Patrick, and Sharpe, I helped organize the Adhesion Society. Today this society's meetings draw participants from all over the world and is the world's foremost organization promoting research and development in the area of adhesion and adhesives.

My graduate students and I have also been active in research relating to the fundamental mechanisms of deformation and failure of materials. While the origin of these research ideas cannot be traced to a specific Gordon Conference, we have attended Gordon Conferences that helped us refine our ideas for papers and proposals. In 1975 I was approached by Sidney J. Green and William Barry about contributing to their proposal for a new GRC called Deformation Mechanisms and Fracture of Polymers and Composites. I was later honored to be elected chair of its 1979 meeting. Many of our research group's ideas on failure in materials either came from or were refined by the papers and discussions heard at these conferences. This conference still meets under the shortened name of Composites.

My attendance at and participation in several dozen Gordon Conferences has been impelled not only by the research incentives outlined above; I have also filled such roles as conference monitor for the Selection and Scheduling (S&S) Committee, board member, and board chair. Gordon Conferences have made me not only a better researcher but also a more interesting and motivating teacher. Participation in these conferences has helped me glean interesting examples of scientific and technical principles, as well as better and more interesting ways of expressing my thoughts.

GRC has also had a profound effect on me through my interactions with the people involved in the organization. One could not help but be impressed by the dedication and managerial acumen of the S&S Committee members, trustees, and GRC staff. I learned much from all of them, particularly the conference directors. During most of my involvement with GRC, Alex Cruickshank was the director, and under his tutelage the conferences grew in number and in national and international prominence. Alex was an effective, firm, and demanding manager, as well as a compassionate and great human being. During my last few years as a board member GRC was directed by Carlyle Storm, who increased office efficiency, brought management into the computer age, increased the number of conferences, and improved and expanded the number of conference sites. I have observed Nancy Gray, GRC's newest director, mostly from afar, but clearly she is, and will continue to be, a great director. She has furthered the growth of the conferences and is exploring ways to enhance overseas operations. I consider all three directors to be not only excellent role models for good management but also valued personal friends.

A FAMILY Reunion

Paul R. Ortiz de Montellano
University of California at San Francisco

Gordon Conferences are like family reunions—complete with the tug of opposing egos, earnest discussions about the future and revisiting of the past, the pleasure of old friends and anticipation of new faces, and gossip about scientific "divorces" and new "marriages." There is also the "Oh, he isn't here again, is he?" syndrome, the planning of new ventures, the nurturing of younger generations, and the farewells of the old. Beyond that, however, Gordon Conferences provide a testing ground for new ideas, a place to match wits with colleagues and competitors and to obtain advice and guidance. All this flows naturally in the context of usually isolated meeting sites and accommodations that often strip individuals of the external trappings of authority (which explains in part why many Gordon Conferences even now prefer the New Hampshire prep-school venues with their too-short beds and down-the-hall baths to sites with hotel-like amenities).

Gordon Conferences are governed democratically, with future GRC chairs selected by popular vote, unlike most serial conferences where a mysteriously appointed group of older statesmen and -women (although rarely) determine who will run the next conference. Although the mechanism for choosing a GRC chair varies from conference to conference, most frequently the slate of nominees for chair is chosen by a committee of previous chairs, with nominations also encouraged from the floor. I remember a meeting of the Agricultural Science Conference when it was trying to reverse a decline in ratings and attendance. Many attendees were from Europe, as agricultural research has a strong presence there, but the nominees proposed by committee were exclusively American. An excellent and personable scientist with academic experience both in the United States and Europe was overlooked in the nomination process. I therefore nominated him from the floor and, not surprisingly given the constituency of the meeting, he was elected as incoming chair. This is not unusual: I have often seen nominations from the floor supersede those offered by the nominating committee. The GRC election process, like all democratic processes, occasionally selects a candidate who turns out not to be a good chair, but the bottom-up process is essential for the continuing renewal of individual Gordon Conferences.

Debate is common at Gordon Research Conferences because the format is specifically set up to promote discussion, which often results in disagreement. I recall one Chemistry and Biology of Tetrapyrroles Conference in which two opposing camps on a contentious issue were asked to make their presentations in a debate format, with each side presenting its case, followed by time for rebuttal. The floor was then opened to the audience to grill the advocates. Was the issue settled? No, but the result of the debate was a crystallization of the possible sources of the disagreement that led eventually to its experimental resolution. This is only one example of the scientific interaction made possible by the intimate nature and flexibility of the GRC format.

Serving a term on the GRC board of trustees was one of the most enjoyable and rewarding experiences of my career. Working with Carl Storm, a highly effective director; staff members who were professional and loyal to the organization; and board members who put aside their personal agendas to work together (at very pleasant meeting venues) made my six years of service gratifying and memorable. Everyone made a strong commitment to the role of GRC as a driving force in international science.

As chair of the search committee that recruited Nancy Ryan Gray to take over the GRC directorship when

Carl Storm retired, I had many interactions with the firm that facilitated the search. The firm's director repeatedly expressed surprise and delight at the ability of the GRC board to focus on issues and to work together for the benefit of the organization (apparently in contrast to many of the searches the firm carried out for other scientific organizations and academic institutions). While I was associated with the organization, its achievements—an aggressive commitment to internationalize GRC by establishing meeting sites in Europe and Asia, funding and construction of GRC's first independent headquarters, the key recruitment of Nancy Gray, financial strengthening, increase of the chairs' fund, and the managed growth of the conference portfolio—certainly did not occur without spirited discussion. But discussion always showed that everyone's concern was focused on the good of the organization.

As a board member and an attendee of conferences at many sites, I was surprised to observe that the ratings of the venues by attendees did not differ greatly from site to site. Some sites were clearly better than others, and the personal comments from the few individuals who bothered to make them reflected this. The overall scores, however, were surprisingly similar, which is a reminder in dealing with a large number of people that individual expectations and requirements differ greatly. For some the proverbial pea-under-the-mattress is a distressing discomfort, while for others not even a boulder in the same location would cause an impression. For some only caviar would bring praise for the GRC meals, but for others even army rations would be happily accepted. Fortunately, enough information filters through the survey process for GRC to carry out a continual pruning of weak sites and replacement with better ones.

The collegial nature of Gordon Conferences brings out the fact that the attendees are human beings and that most are passionate about their work. I can recall the glow of triumph in a young man who had for the first time wowed the audience and tears in the eyes of someone who had just lost out in a scientific race, and even an occasion when a member of the National Academy of the Sciences heaved a can of beer in disgust at a Nobel laureate. Science and the passion that feeds it are the core of the GRC experience.

Transatlantic *Science*

W. Gerhard Pohl

In January 1992 I read about the first Science Education Gordon Research Conference, scheduled for 30 March through 3 April 1992 in Ventura, California. The main reason I applied to participate was because Linus Pauling was the first speaker. I quickly received a letter informing me that approximately ninety people were on the waiting list in front of me. There was no chance I would be admitted.

However, I had included in my application a list of all my activities in the field of science education. As a result I was asked to submit a proposal to offer a similar Gordon Conference in Europe. My proposal for a conference titled New Visualization Technologies for Science Education was accepted in March 1993. To get acquainted with the Gordon Conferences I was invited to participate in the Innovations in the Teaching of College Chemistry Gordon Conference held in Oxnard, California, in January 1994. There I learned a lot about the organization.

John Fackler of Texas A&M University acted as my cochair for the Science Education Conference in Irsee, Germany, and I appreciated his help. In a session called "Visualization of Molecules and Atoms" two Nobel laureates gave lectures: Gerd Binning showed us how solid surfaces can be visualized atom by atom using atomic force microscopy, and Robert Huber discussed the indirect visualization of large molecules by X-ray scattering. Other speakers at the conference talked about electron microscopy and computer graphics as methods to reveal molecular structures and establish structure-function relationships. Ways to use interactive media at science museums were presented by speakers from Cité des Sciences et de l'Industrie in the Parc de la Villette (Paris) and the Deutsches Museum (Munich). Wolfgang M. Heckl, now general director of the Deutsches Museum, showed an impressive film taken with a scanning tunneling microscope of the removal of an atom from a surface layer. Peter Atkins, a chemistry book author, and Ernst Peter Fischer, now a well-known German book author, talked about writing in science.

W. Gerhard Pohl
Austrian Chemical Society

Karlheinz Schwarz
Technical University of Vienna

(continued)

Transatlantic *Science*

The significance of the visualization of molecules has grown since 1994. Various macromolecules are visualized based on X-ray data using computer graphics; a good example is the visualization of a potassium channel of mammalian cell membranes published in *Science* by Roderick MacKinnon's group. Visualization has become a major theme of science education. The Visualization in Science and Education Gordon Conference held at Queen's College in Oxford, England, in 2005 was well attended. Visualization is not only an enormous help for researchers in establishing structure-function relations, but it also allows science educators to explain complex chemical and biological phenomena at a secondary level. Physical molecular models are useful for teaching introductory chemistry, whereas computational graphics are indispensable when dealing with large assemblies of molecules and macromolecules. Progress in the field of visualization techniques continually shows at the Gordon Conference.

Karlheinz Schwarz

In 1994 I chaired a Gordon Research Conference called Phase Transitions in Non-Metallic Solids, which was held at the medieval village of Volterra in the Tuscany region of Italy. It was begun to bring together experts from physics, chemistry, mineralogy, crystallography, mathematics, and materials sciences. In retrospect, the unusual setting has created a very positive atmosphere for fruitful discussions that started many collaborations in this field. For example, one of the speakers, J. Manu Perez-Mato, from Bilbao, Spain, spent a year on sabbatical with my group in Vienna; we have several joint publications that were the result of his expertise in group theory and our competence in first principles calculations.

The second GRC I organized in Europe was called Electron Distribution and Chemical Bonding, held at Queen's College in Oxford in 1998. This conference has occurred triennially for the past thirty years. I have attended most of the meetings, and I must say that these conferences have had a significant impact in the field; it is the interdisciplinary nature of GRC that makes them so important. We brought together experimentalists and theorists—from researchers who focus on instrumentation like synchrotron radiation or area detectors to computational scientists who simulate the properties the experimentalists measure. The combination of the various approaches has led to new insight and brought progress to science.

Several factors make the Gordon Conferences unique. Limiting the number of participants allows time for intensive discussion on unpublished research. The mix of experts from various disciplines—both senior and young scientists—creates an atmosphere in which problems can be openly discussed and constructive criticism can occur. The conference locations are usually far from large cities but are attractive and allow a focus on research without too much distraction. The mandatory requirement that afternoons be kept free from lectures is conducive to informal and fruitful conversation. Afternoon excursions in the surrounding area have led to long-lasting friendships and collaborations. Both Gerhard Pohl and I have seen many joint publications initiated by friendships created at GRC. Finally, alternating conference sites between the United States and Europe has helped strengthen understanding and transatlantic relations among scientists.

Leading the Cutting

Edge

Neil E. Gordon—
"METHOD MUST NOT INTERFERE with
LEARNING"

Leah Shaper
Chemical Heritage Foundation

Neil Elbridge Gordon, Gordon Research Conference founder and science educator, was born to William J. and Ella Mason Gordon on 7 October 1886 in Spafford, New York. After earning a bachelor of philosophy degree in 1911 and a master of arts degree in mathematics in 1912 from Syracuse University, Gordon received a doctorate of philosophy in chemistry in 1917 from Johns Hopkins University (whose chemistry department chair and eventual university president Ira Remsen had pioneered chemistry research and graduate education in the United States). During his career Gordon held several professorships and chaired several chemistry departments, most notably those of Maryland State Agricultural College in College Park, Maryland (1920–1939); Central College in Fayette, Missouri (1936–1942); and Wayne University in Detroit, Michigan (1942–1947). Of significance for the development of the Gordon Conferences he had also occupied the Garvan Chair of Chemical Education at Johns Hopkins from 1928 to 1936.

Gordon's influence on chemical education germinated in the 1920s. Inspired by a paper on undergraduate research presented at a 1921 meeting of the American Chemical Society (ACS), Gordon successfully petitioned that organization to create a new Section (later Division) of Chemical Education, for which he served as secretary and prepared a program that fall. To collect the papers produced by the division, Gordon also launched the *Journal of Chemical Education*. Francis P. Garvan, president of the Chemical Foundation (formed after World War I to purchase German patents for use by the American chemical industry), was so impressed by the journal that he endowed the creation of the Garvan Chair at Johns Hopkins. In that position Gordon developed a fellowship program to support graduate student research, as well as a lecture series designed for uninhibited discussion of research issues; these were the Johns Hopkins University Research Conferences, the first manifestation of what would become GRC, held in Remsen Hall during the summers from 1931 to 1933.

Incorporating innovation into classrooms and professional meetings was of particular interest to Gordon. A chief goal of his unconventional 1927 chemistry textbook, *Introductory Chemistry*, was to incorporate new discoveries into the most recent body of chemical knowledge and introduce them into the classroom. Gordon was specifically concerned that knowledge about the electron's role in the structure of the atom was not being passed on to students, and he noted that because atomic structure formed the basis of all chemical principles, this oversight slowed the learning curve and impeded understanding of chemistry:

> There is inevitably a lag between the discovery of new facts and principles in science and the incorporation of those facts and principles in science into courses of instruction, but it appears to the author that in the application of our present-day knowledge of the structure of matter and in the teaching of general chemistry the lag has been unusually great…. [W]hen all the laws of chemical combination are traceable to the structure of the atom, how can we maintain that we are imparting a knowledge of chemical principles if we neglect the most fundamental principle of all?

Similarly, Gordon conceptualized the conferences at Johns Hopkins as a way of encouraging scientific novelty and progress. These conferences were designed to provide a forum for the discussion of new developments in chemistry and related fields among leading experts and students in a collegial, relaxed, and remote environment that dispelled communication barriers and eased inhibitions, thereby nurturing highly productive brainstorming sessions. Gordon provided just such an atmosphere by moving the conferences in 1934 to Gibson Island, where he encouraged attendees to dress casually, socialize, and partake in recreational activities available in this remote environment. He also guaranteed the confidentiality of conference proceedings and discussion, providing scientists the freedom to discuss novel concepts and experiments well before the publication stage and to get critical feedback from colleagues without the fear of ridicule. "[M]ethod must not interfere with learning," Gordon philosophized in the preface to his textbook.

Frequently described by others as tenacious, persuasive, and resolute, Gordon was considered a highly effective organizer and a relentlessly determined fund-raiser for endeavors that furthered chemical education. In a special fall 1949 issue of the *Record of Chemical Progress* memorializing its editor Neil Gordon, who died in May that year, colleague Otto Reinmuth remembered a man with extraordinary drive:

> In Neil E. Gordon [University of Maryland president G. Forrest Woods] found for chairman of his chemistry department a good executive, a fine organizer, and a promoter of inexhaustible energy. . . .
>
> He thoroughly sold himself on a project before he set out to sell others, and he never tried to feather his own nest. His obvious sincerity and selflessness, coupled with his extreme artlessness, doubtless disarmed many who were used to being high-pressured by smooth operators. He was optimistic to the point of blindness, and the continual dropping that weareth away the hardest stone had lessons in persistency to learn from him.

> . . . Whatever aspirations toward scientific scholarship he may once have entertained, he early recognized his own peculiar abilities as an organizer and promoter, and subordinated all else to the furtherance of projects which he felt would advance the cause of chemistry and chemical education.

The conferences that had already begun in the late 1920s and were formalized at Johns Hopkins in 1931 grew steadily in the mid-1930s. Gordon, however, wanted to move on to other endeavors. By persuading the American Association for the Advancement of Science (AAAS) to administer the conferences in 1937, he took a critical step toward their permanence. In a November 1939 letter to AAAS executive secretary Forest R. Moulton, Gordon wrote:

> This summer session needs to take a little more definite form of organization. I should like to bring it to such a status this summer that it could be continued as a permanent function of the association. I should like to try to put the whole thing on a paying basis, and extend it over at least six weeks of the summer months.

To do this, Gordon also felt it necessary to establish leadership of the conferences and suggested that he be named their director.

> This would mean that I should have to have power to go ahead and organize it, selecting a chairman for each week. I realize that if I am to do this, I must do it so as to guarantee no expense to the section or to the society. I believe it might be helpful in this work, if I were made director of the summer session. It is beginning to become a proposition to handle it as secretary to the section, without some other title, for I shall already have under me at least six chairmen to carry out this work.

Yet in the April 1948 dedication ceremony that officially named the Gordon Research Conferences in honor of their founder, Gordon accepted the honor modestly and characteristically deferred credit for the conferences' success.

(continued)

Neil E. Gordon—
"Method Must Not Interfere with Learning"

I am accepting this honor in the name of the men who have participated in the conferences because of their cooperation, and hope that I may have the privilege of working with them in the future to do what little I can so that the conferences may be much more successful during the next ten years than they have been during the past ten years.

Gordon's retirement as director of the conferences in 1947 did not put an end to his efforts to improve the state of chemical education and professional development opportunities for chemical researchers and educators. As chair of the Wayne University chemistry department, Gordon established the university's first doctoral program. He also succeeded in creating the Kresge-Hooker Library, a large collection of chemical texts originally compiled by the industrialist Samuel C. Hooker. During this time Gordon was also busy directing a lecture series called Frontiers in Chemistry as part of an academic course at Wayne. The series was soon published in a two-volume set called *Advancing Fronts in Chemistry*. The volumes' prefaces by Sumner Twiss and Wendell Powers echo the principles established by Neil Gordon for the conferences at Johns Hopkins:

> There is a special need for the correlation of recent and diverse experimental information.... This series of lectures is, therefore, an attempt at such a correlation, presenting as it does the unified concepts of certain phases of the polymerization problem as they are understood and interpreted by various experts in the field. (Twiss, 1945)

> It has been [Gordon's] aim in securing lectures for this course, to invite specialists from the industrial as well as the academic chemical field, who are actively engaged in the development of certain fundamental branches of chemistry. (Powers, 1946)

Behind the early years of the conferences was the support and idealism of Neil Gordon, in whom intersected a powerful combination of sheer determination, fund-raising and networking skills, and a passion for chemical education. This concurrence of traits provided an incredible momentum to the Gordon Research Conferences that has carried them through to their seventy-fifth year and beyond.

The American Association for the Advancement of Science (AAAS) has been a long-standing partner of the Gordon Research Conferences (GRC). In June 1937, at the AAAS meeting in Denver, Neil E. Gordon approached Forest R. Moulton, permanent secretary of AAAS, with the idea that the association sponsor a program of small research conferences on chemistry that he had led since 1931. The AAAS executive committee voted to authorize the conferences, provided that the association would not incur financial liabilities. Moulton recounted that "this action was reported to Dr. Gordon somewhat hesitantly, because it amounted to saying that if the conferences should be successful, the association would profit by them, and that if they should be unsuccessful, Dr. Gordon would be the only sufferer." Gordon accepted the conditions, and the conferences indeed proved successful.

AAAS sponsored GRC as a formal program, sharing in the administrative stresses that came with the increasingly complex organizational venture. AAAS established an advising policy committee, composed primarily of industry scientists, with Gordon as chair and Moulton representing AAAS. By the summer of 1941 the conferences had grown from two to eight topical programs, offered during the summers on Gibson Island in the Chesapeake Bay near Baltimore. Sixty selected participants took part in each five-day AAAS–Gibson Island Conference program. At least thirty-three chemical companies contributed $1,000 apiece toward AAAS's purchase and improvement of a large house on the island, soon renamed the Neil E. Gordon House, that provided extra accommodations for program attendees and a porch on which to hold program meetings. Further efforts to expand the conferences created friction with members of the nearby Gibson Island Club, who objected to over-

AAAS
and the
Gordon Research Conferences

crowding of the club's lounging facilities and the "somewhat informal attire" of the scientists, who often failed to wear jackets in areas where the club required them. At the recommendation of the GRC's management committee, the 1947 summer sessions were moved to Colby Junior College in New Hampshire. In 1948 the program was renamed the Gordon Research Conferences, and AAAS sold the Gibson Island property.

In the early years of the association's partnership with the conferences the AAAS executive committee authorized final approval of conference management decisions. Moulton kept AAAS officers apprised of conference activities. In 1956 the management committee incorporated GRC in New Hampshire, which provided the newly independent organization tax-exempt educational status and relieved the association of its duties as trustee of the conference funds. The organization's new constitution and bylaws established the GRC board of trustees, formed from the former management committee. Approval from the AAAS board of directors continued to be required for amendments to the constitution, but GRC, which managed its own affairs, rarely invoked this provision and excised it in 1980. Today, the AAAS chief executive officer still serves as a member of the GRC board, and the AAAS publication *Science* continues to publish conference program announcements twice a year.

The conferences' style and content has affected science and its applications tremendously. Early conference organizers maintained that the "objective" of GRC was "to sharpen [participants'] wits; not to obtain data." GRC's off-the-record policy—which discourages taking detailed notes, prohibits photographing slides or recording presentations, and for some time banned attendance by reporters —encourages in-depth exchange among peers. Philip Abelson, former editor of *Science*, noted that the GRC provides "secluded spots where participants are free of distractions and the presence of nonscientists." This format has fostered discussion about GRC's broad effects on science and society. Former AAAS executive officer Dael Wolfe credited the 1973 Nucleic Acids GRC as an influential factor in the instigation of debate and subsequent conferences regarding the potentials and hazards of recombinant-DNA research and ultimately the National Institutes of Health guidelines for its safe practice. Richard S. Nicholson, another former AAAS executive officer, posited that "GRC's cumulative contribution to communication and the advancement of science is enormous." With more than 350 Gordon Research Conferences on topics in biology, physics, chemistry, and science education and policy, the conferences' special format and growing international character poise the organization to continue as a major contributor to the advancement of science around the world.

Amy Crumpton and Alan Leshner
American Association for the Advancement of Science

Dr. Gordon's Serious

Nature is a vast, diversified farm and factory. Its growth and manufacturing processes have fascinated man for thousands of years, and he has tried repeatedly to mimic what he has observed. Sometimes he has succeeded to a degree that has astonished himself. Most people are not yet aware that there is a new world to be discerned out on the far frontier of science. But research pioneers in the world's institutions of learning and the far-flung into the

Improvements in carbon black, which the carbon black markers laughed at when they originally heard the notion at a Gordon Conference, have transformed the auto tire from a 3,500-mile frailty to a puncture-proof reliability that can be made to last 100,000 miles. And now, after having learned how to start and stop the growth of polymers, scientists gathered for the Gordon meetings this year have finally reported discovery of the precise mechanics by which Nature controls the growth ... ered for the Gordon

W. George Parks
Director Emeritus, GRC

Nature is a vast, diversified farm and factory. Its growth and manufacturing processes have fascinated man for thousands of years, and he has tried repeatedly to mimic what he has observed. Sometimes he has succeeded to a degree that has astonished himself.

Most people are not yet aware that there is a new world to be discerned out on the far frontier of science. But research pioneers in the world's institutions of learning and the far-flung empires of industry are probing and peering into the misty fringes of a wholly new and probably more fantastic era of discovery than any other that man has known. The shape of that era is already visible on the horizon that stretches out before the Gordon Research Conferences.

Nobel Prize–winners of world renown ... meet and mingle here, an unbroken week at a stretch, with bright young apprentices in thirty-six different realms of science during the summer. They go about the three host campuses, spaced thirty-odd miles apart—Colby Junior College, the New Hampton School, Kimball Union Academy—without neckties, in gaudy open-collared short-sleeved shirts, shorts, and sneakers, arguing from dawn to dawn, secure in the Gordon Conference guarantee that they will never be quoted and that anyone they encounter in their perambulations will be a fellow scientist intent on common problems.

The Italian-born father of A-power, the late Enrico Fermi, described his now-famous slow-neutron process of atomic disintegration to a Gordon Research group in 1936, a decade before the A-bomb. Early work on the vitamins was advanced by Gordon Conference discussions a quarter-century ago.

Improvements in carbon black, which the carbon black makers laughed at when they originally heard the notion at a Gordon Conference, have transformed the auto tire from a 3,500-mile frailty to a puncture-proof reliability that can be made to last 100,000 miles.

And now, after having learned how to start and stop the growth of polymers, scientists gathered for the Gordon meetings this year have finally reported discovery of the precise mechanics by which Nature controls that growth at ordinary "room temperature" in the open air. The method is the same one that regulates the growth of hair, grass, and trees. From this it seems inevitable that man will have the cheapest clothing in history, along with plastics of incredible strength, revolutionary types of films and glass, and probably subtle new medicines derived from the enzymes which keep us all alive and growing.

"Anyone would be honored to be invited to talk to a Gordon Conference on any subject," a British atomic scholar told the meeting where he appeared this summer. Like sentiments have been expressed to us by scientists in England, in Europe, in Africa, in India and Japan, in Australia, and in South America. At no other place than the Gordon Research Meetings are conferees afforded the leisure for camaraderie which exists in these graceful, green hills. From Sunday morning to Friday night there are only nine formal sessions, and each of the nine occupies about two hours. Roughly one hour is staccato with discussion, and the speaker in the first hour can be interrupted at will by any of his listeners. Argument goes on throughout the rest of the week, while the conferees climb mountains, swim and sail the lakes, explore nearby caves, watch for birds in the early mornings, bang sporadically on an old piano and sing songs together in the middle of the night, play in every conceivable fashion from tennis and

hinkers

Editors' note: The late W. George Parks was the director of the Gordon Research Conferences from 1947 to 1968. The following abridged essay has been reprinted with permission from the 4 August 1956 issue of *Saturday Review* (pp. 42–49). Permission was graciously donated by Bob Guccione, who congratulates the Gordon Research Conferences on its 75th anniversary.

golf to ping-pong and bridge, from string quartets and square dances to table-stakes poker.

The scientists growl each other awake at breakfast, swap tablemates repeatedly to rotate shop talk at lunch and dinner, lie in wait for famous companions after meetings, and even accompany departing speakers to the airport thirty-four miles away in order to get in a few more licks during the taxi ride.

The Gordon Conferences have been so secluded that they have gone on for twenty-five years without being known to the lay public or even to many scientists. Strictly an invitation affair, they are rigorously limited in size to assure the utmost communication of advanced ideas. No other meetings are allowed on the host campuses throughout the summer, so that any conferee may talk freely to anyone he meets without fear of violated confidence. All discussion is off the record. Reporters are barred.

[The] late Dr. Neil E. Gordon, the Johns Hopkins University chemistry professor ... realize[d] far better than many more brilliant and gifted contemporaries that growing specialization was gradually cutting science up into small, isolated bits and that something had to be done to fit those fragments to the increasingly complex problems of human society.

It was the fashion then to be concerned with applied science, with the practical use of past discovery. Dr. Gordon moved in the opposite direction, out to the frontier of the future.

Formal papers were discouraged from the very beginning. Open, free discussion was the purpose of the meetings, and the subject matter was circumscribed to research in progress, incomplete experiments, new theories not yet boiled down onto paper for publication. Gibson Island became a scientific testing ground, a place where daring young minds could throw offbeat ideas at older heads crammed with experience and skepticism. Only two meetings were held each day, one in mid-morning, one after dinner at night. The rest of the time was spent in sporadic discussion on the golf course, on the tennis courts, in the swimming pool, on the bayshore with fishing pole in hand, or out on the water under sail. The trumpet vines, the mockingbirds, the osprey's nests, and the mint juleps were just about as far from the conventional conception of a test tube as can readily be imagined. Yet the island was in constant ferment with original thought. Charles Allen Thomas, now president of the Monsanto Chemical Company, recalls how he as a "struggling young research chemist" was invited to a session and returned to his lab "with renewed vigor ... to my work."

Dr. Gordon's rule for dress on Gibson Island was "No jacket, no necktie, shorts if you want to wear them." This shocked the staid members of the Gibson Island Club, even after the Gordon Conferences moved from the club lounge to the boathouse.... A climax arrived when one conferee overheard an exchange between two Baltimore aristocrats ensconced on the club verandah, in rocking chairs. "Who are those people who have invaded our island?" the one lady inquired. "Oh," the other replied with a sniff, "they're Doctor Gordon's serious thinkers."

The seriousness of the thinking has not been lessened since the Gordon Conferences shifted from the muggy Chesapeake to the cool, pond-dotted undulations of New Hampshire in 1947. But the natives of the White Mountain foothills are proud of it. "When," they ask me each spring, "will the scientists come this year?"

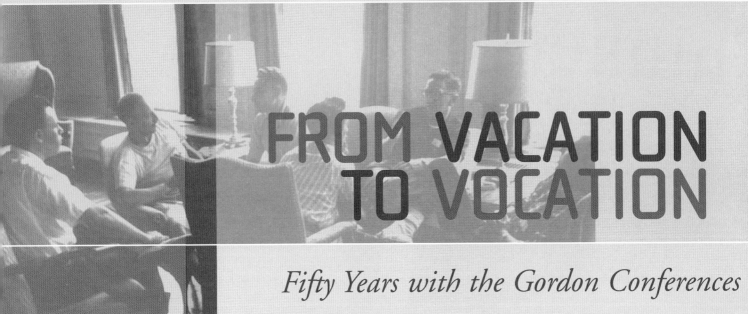

FROM VACATION TO VOCATION

Fifty Years with the Gordon Conferences

Alexander M. Cruickshank
Director Emeritus, GRC

New Years Day 1947 my wife Irene and I were having dinner with Dr. and Mrs. George Parks, when he asked, "How would you like to vacation in the mountains of New Hampshire for ten weeks this summer, all expenses paid?" At that time I was a young chemistry instructor at Rhode Island State College (renamed University of Rhode Island in 1951), and Irene was secretary to Dr. Parks who was a full professor. He presented his offer after Dr. George Calingaert of Ethyl Corporation had written a letter asking him to direct the Gibson Island Research Conferences during the summer of 1947. The conferences' founder, Dr. Neil Gordon, had resigned the year before. Irene and I hesitated until Parks told us he would not take the job unless we agreed. It was decided: Irene became the secretary and treasurer, and I assisted wherever necessary the first summer and in 1948 became assistant to the director during the summer sessions.

Not only had Gordon resigned in 1946, but the Gibson Island Research Conferences had also outgrown their meeting facilities. Led by Dr. Calingaert, a management committee of the American Association for the Advancement of Science set a foundation for the conferences that maintained their success and sought a new meeting site. The committee also included Drs. Kenneth Hickman, Lawrence Flett, Sereck Fox, Dean Burke, and A. Marshall. On their way back from an unfruitful visit to Dartmouth College (there were too many other summer activities on campus), they passed by Colby Junior College (renamed Colby-Sawyer College in 1975) and stopped to talk to its president, H. Leslie Sawyer. Colby Junior was a perfect fit: it was cool, secluded, and otherwise closed during the summer. As the conferences were no longer on Gibson Island, they were temporarily renamed the Chemical Research Conferences in 1947 until they became the Gordon

Research Conferences in 1948 and were incorporated in 1956. These arrangements kept the conferences going and laid the groundwork for their growth.

In the early years either Dr. Parks or I would open the meeting on Monday morning. There was only one conference a week then, and my job was to keep the operation moving. Initially I would run the slide projector, deal with the attendees, and handle on-site logistics. I also helped Irene in the office whenever I could, since keeping the financial records in order was a big job. In addition, I kept statistics on the conferences, including the number of participants and their affiliations with academia, government, or industry. In the 1940s conferees came largely from industrial research labs, partly because chemical, petroleum, and pharmaceutical firms sponsored the conferences and partly because that was where a lot of leading-edge science was going on. By the mid-1950s the government was sponsoring more scientific work, and participation by academics in the Gordon Conferences increased rapidly.

Each conference always had lectures and discussion. The discussion was very important; in fact, we had monitors that reported on the amount of discussion (and even timed it). A. Marshall of General Electric was the driving force behind creating the Selection and Scheduling Committee in 1958 to review the previous year's conferences and applications for new conferences. The group met in October, sometimes for two days, and went through stacks of monitors' reports. This ensured the quality of the conferences: if a conference did not do well or if there was very little discussion, the committee did not hesitate to terminate it. Conferences also monitored themselves. If a conferee was observed to be merely taking notes and not contributing to the discussion, he or she might not be accepted the next time the conference met. Attendees quickly learned they had to participate.

Scientific discussions are the most amazing thing about the conferences. People on the floor would argue with each other. In fact, they came to Gordon Conferences to challenge each other and defend their own science. If a conferee had evidence contrary to someone else's presentation, pretty often they would have a slide of it in their back pocket. They were just waiting for the opportunity to present new findings. I recall Professor Herman Mark bringing equipment and supplies to present polymer experiments in front of the conferees.

The Gordon Conferences promoted a level playing field: we often had a mix of Nobel laureates and young scientists in attendance. Well-known chemists like Professor Paul Emmett (Catalysis Conference) were great with all the attendees. Dr. Emmett would sit under a tree every afternoon surrounded by attendees asking questions. This was source material for them. At all the conferences the attendees would spend their entire afternoon just exchanging ideas. Dr. Parks, Irene, and I enjoyed seeing those one-on-one relationships develop and did everything we could to foster them. Afternoon free time and meals at round tables were especially nourishing for discussion with table linens used to write on.

Part of what made my career with GRC special was getting to know the conferees firsthand. The afternoons were free, and so I would help the scientists find something to do, or I would take them down to the local swimming hole, where a lot of science was discussed over a raft. We knew 30 to 40 percent of the people attending on a first-name basis, so each conference was like a family reunion. We would register them and assign them a room, and then they would gather around in the yard. Dr. Parks and I would go out and talk with the conferees. In the evening we used to stand around on the campus road after dinner and answer their many questions. As the conferences expanded, I continued to make a point of visiting the conference chairs at all twelve sites to review their specific agendas.

When I became director in 1968, we adjusted the conference fees and put surplus money into a special fund. Each chairperson was then given five hundred dollars to use for speaker expenses, travel, or registration fees. Eventually each conference was getting three thousand dollars. In many cases government agencies would match a conference chair's fund. We also applied for federal grants without an overhead charge for individual conferences and convinced chairs to write their own grant proposals, which GRC administered. Throughout the years GRC staff has always worked diligently to assure the successful operation of the conferences.

Even before I was made director, the conferences were branching out geographically. The first group to meet in California was the Polymers Conference in 1963. Professor Herman Mark had a large contingency of polymer scientists on the West Coast, one of whom suggested that the conference meet at the Miramar Hotel near Santa Barbara. It was isolated, on the oceanfront, and the grounds were lush and gorgeous. The conferees could go out, sit, and talk to colleagues under palm trees, surrounded by flower gardens with the two outdoor swimming pools— the site most conducive to lots of afternoon discussions.

We always worked to bring in scientists from around the world. Even during the cold war we included scientists from Eastern Bloc nations. Shortly before I retired, we held our first overseas conferences, which came about because some scientists wondered why they always had to come to the United States to go to a Gordon Conference. After the GRC trustees had a joint meeting with several European scientific organizations in Irsee, Germany, we scheduled the first overseas GRC in 1990 in Volterra, Italy, and in 1991 in Irsee, Germany. In 1993 I toured England with trustee Dr. Leila Diamond to explore a new meeting site that opened at Queen's College, Oxford, in 1996 after I retired.

The board of trustees has played a critical role over time by constantly questioning whether the organization was expanding too quickly, whether conferences were becoming less personal as they grew, and whether we were on the leading edge of new scientific fields. Throughout the years GRC has held to the ideals laid out by Gordon that scientific fields flourish only through good communication in an atmosphere of collegial challenge. Gordon's zeal, vision, and understanding of the importance of communication brought the conferences into being, and I had the pleasure of watching them grow, expand, and mature. Unlike other organizations, however, I believe that the Gordon Research Conferences will continue to prosper and expand on the incredible enthusiasm and energy of the participants and their dedicated chairpersons.

Irene and I were most fortunate, thanks to Dr. Parks, to be a part of the Gordon Research Conferences "family," and to say we enjoyed our work would be an understatement. The early years required long hours and much trial and error, but they were still lots of fun. Many conferees became close, lifelong friends, and we continue to cherish very much the fond memories and experiences we had over the years. Every conference, every conferee, every week, every year, was exciting and most rewarding: we were lucky.

A Fellowship of

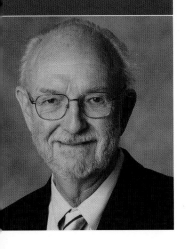

Carlyle B. Storm
Director Emeritus, GRC

In 1667 Queen Christina of Sweden praised chemistry as "a beautiful science, and the key which opens all treasures, giving health, glory and wisdom." Even as the constitution of the discipline has changed enormously, this working definition of chemistry still applies today. For example, atomic and molecular science and technologies have expanded over the years and now cover conventional areas of biology, physics, materials science, and engineering. Originating in discussions held at the chemistry department of Johns Hopkins University, the Gordon Research Conferences have followed this evolution of research interest as science has moved toward an atomic and molecular understanding of material and processes. Scientists who regularly gather at Gordon Conferences enjoy the optimistic charge of Queen Christina and the current pleasures of exploring the frontiers of science.

In my own career I have profited from and enjoyed participating in Gordon Conferences immensely. I have participated in the Chemistry and Biology of Tetrapyrroles, Metals in Biology, Organic Geochemistry, Isotopes in Chemical and Biological Sciences, Inorganic Chemistry, and Energetic Materials (EM) conferences. I was the founding chair of the EM Conference while on the staff of Los Alamos National Laboratory. In the mid-1980s Dick Miller, from the Office of Naval Research, and I arranged a meeting at the New Mexico Institute of Mining and Technology for research scientists from industry, academia, and national and military laboratories to discuss energetic materials R&D. At the end of the meeting many participants enthusiastically observed that the meeting was "just like a GRC." I was given the task of approaching

GRC about establishing a regular meeting on energetic materials. Our proposal was approved and the EM GRC had its first session in 1986.

Energetic Materials is an example of how a Gordon Conference can help bring together R&D participants from a variety of fields. Energetic materials covers a diverse group of materials, including high-performance high explosives, insensitive high explosives, propellants, emulsion explosives, pyrotechnics, and explosive welding techniques. The chemistry and physics of the materials are specialized, and the participants come from a broad international group representing industry, military applications, space exploration, university-based research, and other communities. These groups can be quite competitive and did not previously have a common venue to discuss fundamental research issues in the chemistry, physics, and materials science of these materials. The EM Gordon Conference has provided a common interface where basic problems can be discussed and relationships established. When the EM Conference was established, I was doing mostly program management and technology transfer at Los Alamos. GRC allowed me to make valuable contacts and to see what was taking place across the energetic materials field.

When I moved from Los Alamos to GRC as its director in 1993, a number of people suggested that the organization was not broken and that I not try to fix it. Put differently, GRC is a valuable asset to the scientific community, and no radical changes were needed in its organization or operation. I have followed that advice to a considerable extent. While working with the board of

Science

directors, the Selection and Scheduling Committee, and the council, GRC has grown at a steady, but manageable, rate. Business administration is now Internet-based; headquarters are housed in a new building; international Gordon Conferences are a success; and we cover a more diverse set of topics in science and technology, including the social impacts of those topics, than ever before.

The success of the GRC format lies in the inherent nature of group dynamics. Scientists tend to be synthetic thinkers; that is, they are people who are always on the lookout for new connections. A Gordon Conference provides many opportunities to view a problem from new vantage points, to seek personal connections, and to learn about new experimental methods and theoretical constructs.

Further, speculation is inherent to the development of a new idea or theory. Productive speculation requires an approachable climate of free and open idea exchange. GRC is famous for providing an opportunity for graduate students to meet and debate with the most senior members of a scientific community. Instant peer review is always available.

Despite misconceptions that genuine innovators are loners and do not need the social reinforcements that many crave, great innovations often come from group interactions. Whether in art, politics, philosophy, or science, innovation arises from social interaction—conversation, validation, the intimacy of proximity, and the look in a listener's eye that tells the creator that he or she is on to something.

A cluster of people will come up with a position or decision far more extreme than that proposed by an individual. The flux and constant reassessment inherent in group decision making moves a group on to new possibilities, weeds out the weak ideas, and generates radically different approaches.

Finally, spending a week in a comfortable, rural setting with friends who share a strong interest in a compelling, fascinating subject is just good plain fun. The open and relaxed format, often termed "GRC summer camp," brings people together, lowers the barriers to communication, and builds good group interactions.

When I became director of GRC, I inherited the efforts of Neil Gordon, George Parks, Alex Cruickshank, and the many dedicated volunteers who provided guidance and governance for the organization. I have enjoyed working with the GRC governing bodies immensely, the two thousand or more conference chairs over the past decade, and the many thousands of conferees I have met and often shared a meal or drink with. They have well represented my view of GRC, that it is a "fellowship of science."

Finally, I would like to express my appreciation to the GRC staff for their continuing efficient and good-natured efforts and the excellent service they have provided to the many thousands of conferees. They provide the infrastructure that makes the whole thing work.

Before *the* Pros Arrive

Gordon-Kenan Graduate Research Seminars

Ruben G. Carbonell and Rajeev Narayan
William R. Kenan, Jr., Institute for Engineering, Technology & Science

The William R. Kenan, Jr., Institute for Engineering, Technology, and Science at North Carolina State University (Kenan Institute) proudly continues its collaboration with GRC at its seventy-fifth anniversary to foster scientific innovation through the productive exchange of ideas. The institute is especially gratified to be able to collaborate with GRC to foster scientific innovation. In February 2001 the Kenan Institute and GRC began a partnership to enable new opportunities for young researchers and scientists to participate in Gordon Conferences. This partnership resulted in the Gordon-Kenan Graduate Research Seminars, which have allowed advanced graduate students, postdocs, and established scientists to participate jointly in a wide range of conferences and share in the GRC experiences.

As the Gordon-Kenan collaboration continues, diverse topical areas are being explored by Graduate Research Seminars (GRS) such as Bioinorganic Chemistry, Molecular Energy Transfer, Protein Folding Dynamics, Bioanalytical Sensors, and Origins of Life. The Kenan Institute also encourages faculty from North Carolina State University and others to submit proposals to GRC for seminars in such areas as innovation management.

The Gordon-Kenan Graduate Research Seminars provide the opportunity for younger scientists to interact with one another at a "mini-GRC." Generally, this allows younger scientists and students to warm up, get to know one another, and learn more about the caliber of their peers' research in a "safe" setting before the pros arrive. Participants in a GRS can also learn more about the cutting-edge research in a given field from the graduate students who are carrying it out, and they gain experience presenting research in the setting of a scientific conference. First-rate presentations on topics on the forefront of a given discipline not published in the scientific literature are expected at Graduate Research Seminars. The opportunity to network and have informal discussion on a personal and a professional level with top young scientists in the early part of their careers is an important element of a GRS.

The Kenan Institute's mission is to develop partnerships in basic research, education, commercialization, and public outreach with individuals and organizations dedicated to the advancement of science, engineering, and technology, with partnerships acting as forces toward improving the economic and social well-being of the state, nation, and world. The unique partnership between the Kenan Institute and GRC leverages the great traditions of both institutions in an innovative manner, bringing together diverse groups of scientists from around the world who share their knowledge and forge important collaborative relationships.

The Kenan Institute's collaboration with GRC is supported by a generous financial contribution from the William R. Kenan, Jr., Fund for Engineering, Technology, and Science. Kenan was a bright, young scientist in his own right who worked closely with Frances Venable in the late 1890s and helped make significant contributions in basic and applied research and the commercialization of technology. In the same spirit the Gordon-Kenan collaboration seeks to enable young and experienced scientists to work together and generate new discoveries beneficial to society. In his will Kenan wrote, "I have always believed firmly that a good education is the most cherished gift an individual can receive." The Gordon-Kenan collaboration continues to honor Kenan's commitment to excellence in education by providing young scientists with the very best opportunities to share their knowledge and learn from one another and from more experienced scientists.

Appendix

GRC Directors

Nancy Ryan Gray
2003–present

Carlyle B. Storm
1993–2003

Alexander M. Cruickshank
1947–1993 (Director and Assistant Director)

Irene Cruickshank (Secretary and Treasurer from 1947 to 1955) and Alexander M. Cruickshank

GRC Management

Nancy Ryan Gray
Darlene Graveline
Gerri Miceli
Jim David
Suzanne Tucker

GRC Staff

Nancy Ryan Gray, *Director*
Darlene Graveline, *Office and Operations Manager*
Gerri Miceli, *Program Manager*
Holly Tobin, *Grants and Development Assistant*
Michelle Knapp, *Administrative Assistant*
Pamela Potter, *Administrative Assistant*
Jim David, *IT Manager*
Jeff Carroll, *IT Support and Webmaster*
Brad Sharkey, *IT Support*
Lynne Kindle, *Conference Coordinator*
Maril Cappelli, *Conference Coordinator*
Cathy McGinn, *Conference Coordinator* (not pictured)
Rebecca Newman, *Conference Coordinator*
Suzanne Tucker, *Financial Operations Manager* (not pictured)
Brenda Goodness, *Accounts Receivable*
Beverly Mason, *Grants and Contracts*
Deborah Luz, *Grants and Contracts*
Lynn Irwin, *Grants and Contracts*
Beth Wheeler, *Registration Processing*
Paul Elkins, *Summer Site Coordinator*
Adam Zuber, *Winter 2006 Site Coordinator*

ABOUT THE GRC HISTORY project

This commemorative publication is the culmination of a four-year collaborative project between the Gordon Research Conferences and the Chemical Heritage Foundation to collect and preserve the GRC story, to explore GRC's rich history and incredible growth, to make known its integral role in scientific progress, and to celebrate its seventy-fifth anniversary.

The project began with the transfer, processing, and preservation of GRC's historical records, which are now available at CHF to scientists and historians. Next the team (shown in photo from left to right, Leah Shaper, Arthur A. Daemmrich, Nancy Ryan Gray, and Pamela Potter) produced a poster, "History and Evolution of the Gordon Research Conferences," which documents GRC's growth and role in stimulating scientific innovation. GRC's seventy-fifth anniversary year features the release of this publication and the launch of the Frontiers of Science Web site, an online overview of GRC history showcasing its growth, personal narratives and recollections, oral histories with past GRC leaders, and reference tools. The site will also provide the opportunity for conference participants to tell their stories.

Historical research, analysis, and writing for the project drew on the GRC archives, other primary and secondary sources, oral history interviews, quantitative databases of attendee demographics, and qualitative surveys of participants. Essayists in this publication were selected through consultation with the history project advisory committee to reflect the breadth and depth of participants at Gordon Conferences.

ACKNOWLEDGMENTS

Just as attending a Gordon Research Conference is a unique educational experience for thousands of scientists each year, editing a volume with eighty contributors was a remarkable undertaking. We are grateful to everyone at the Gordon Research Conferences and the Chemical Heritage Foundation for their help, especially Darlene Graveline, office and operations manager, and Pamela Potter, administrative assistant, at GRC and Chi Chan, program assistant, and Andrew Mangravite, archivist, at CHF. The CHF communications and marketing area deserves special recognition, particularly Patricia Wieland, editorial consultant, and Shelley Wilks Geehr, director of communications and marketing.

GRC History Project Advisory Group

The project was guided from its inception by an advisory group. We are most grateful to its members for offering counsel and continuity, for identifying and recruiting essayists, and for their overall enthusiasm. The members were Paul Anderson, Alexander M. Cruickshank, James R. Florini, Frances A. Houle, Gerri Miceli, Mark Ratner (chair), Carlyle B. Storm, Joan Selverstone Valentine, and Barbara K. Vonderhaar.

This publication was partially underwritten through a generous grant from Pfizer, Inc.